Methodological Approaches for Workplace Research and Management

This book explores a wide range of methodological approaches to examining various forms of workplace physical environments. It focuses on pressing questions regarding the relationship between the spatial component of the workplace, including its progressive hybridisation with other physical and virtual places, and its users, be they public organisations, private companies, or start-up businesses and solopreneurs.

International contributors address a range of methods that are applicable both in research and practice to confront the most cutting-edge workplace-related issues. The assumption is that work has been changing, thanks to the virtualisation of many activities, and that homeworking and hybrid working modes are expected to increase significantly after Covid-19. Thus, spaces hosting work need to adapt accordingly. Researchers and practitioners have been struggling to determine how much space will be needed by companies, what kind of space will better host different work activities, which workers are more suited for working from home, and which instead are more productive if they have an office-based working arrangement. The necessary evolution of the office should follow evidence-based decisions on the abovementioned matters, which are only possible through rigorous investigations. This volume aims to support these investigations, which call for inventive applications of qualitative and quantitative methodologies. By doing so the book will encourage solid practices and thorough research agendas in workplace design, management, and use.

Contributions come from different disciplines, including facilities management, real estate management, psychology, design, architecture, sociology, and organisation studies. Chapters highlight the importance of appropriate methodologies, borrowed from different fields, in addressing contemporary questions and developments in workplaces. By analysing the challenges and opportunities for conducting rigorous research in different workplace settings, this book will be critical reading for both academics and students, as well as for decision-makers and professionals who deal with workplace design and management.

Chiara Tagliaro works as a fixed-term researcher and consultant in the Real Estate Center – Department of Architecture, Built Environment, and Construction Engineering at Politecnico di Milano. Her research interests concern the design, management, and use of workplaces, collaborative work practices, and the digitalisation of the real estate sector. She has been involved in two EU-funded projects around these topics: the COST Action CA18214 "The geography of new working spaces and the impact on the periphery"; and the ERASMUS+ Strategic Partnership 2018-1-HU01-KA203-047744 "An innovative approach in workplace management education". She has coordinated the Summer School on Workplace Management since its first edition in 2018. She organised the third TWR conference in 2022 in Milan, Italy.

Marko Orel is an organisational sociologist who works as an assistant professor and a head of the Centre for Workplace Research (CWER) at Prague University of Economics and Business. He specialises in exploring the changing nature of the workplace and the transformation of work and work-related processes. In addition to that, Marko is currently exploring experimental, qualitative research methodologies. He recently guest-edited a special issue on workplace transformation at Emerald's *Journal of Corporate Real Estate*, edited a volume on flexible workplaces that Springer Nature has published, and has published several chapters and research papers in journals such as *World Leisure Journal, Mobile Networks & Applications, Review of Managerial Science*, and others.

Ying Hua is an Associate Professor (tenured) in the Cornell Department of Human Centred Design – College of Human Ecology and is the Director of the Cornell China Center. She is a member of the graduate field and advisor in the Cornell Baker Program in Real Estate, a faculty fellow of the Atkinson Center for a Sustainable Future and the Cornell Institute of Healthy Futures, Vice-Chair of the Faculty Advisory Board of the Atkinson Center, and core faculty of the Cornell East Asia Program. Dr Hua leads the International Workplace Studies Program (IWSP) with both research and consulting work. Her research addresses design and management challenges across multiple phases of project life-cycle, with a particular interest in future work and workplace in the US, Japan, and China. She has a track record of 17 papers indexed in Scopus (h-index 9), and has been a frequent speaker at international conferences and high-profile events.

Transdisciplinary Workplace Research and Management
Series Editors: Rianne Appel-Meulenbroek and Vitalija Danivska

A Handbook of Theories on Designing Alignment Between People and the Office Environment
Edited by Rianne Appel-Meulenbroek and Vitalija Danivska

A Handbook of Management Theories and Models for Office Environments and Services
Edited by Rianne Appel-Meulenbroek and Vitalija Danivska

Methodological Approaches for Workplace Research and Management
Edited by Chiara Tagliaro, Marko Orel and Ying Hua

Methodological Approaches for Workplace Research and Management

Edited by Chiara Tagliaro, Marko Orel and Ying Hua

Routledge
Taylor & Francis Group
LONDON AND NEW YORK

Designed cover image: © Getty Images

First published 2023
by Routledge
4 Park Square, Milton Park, Abingdon, Oxon OX14 4RN

and by Routledge
605 Third Avenue, New York, NY 10158

Routledge is an imprint of the Taylor & Francis Group, an informa business

British Library Cataloguing-in-Publication Data
A catalogue record for this book is available from the British Library

Library of Congress Cataloging-in-Publication Data
Names: Tagliaro, Chiara, editor. | Orel, Marko, 1986- editor. | Hua, Ying
(Writer on workplace studies)
Title: Methodological approaches for workplace research and management/
edited by Chiara Tagliaro, Marko Orel, and Ying Hua.
Description: Abingdon, Oxon; New York, NY: Routledge, 2023. |
Series: Transdisciplinary workplace research and management |
Includes bibliographical references and index.
Identifiers: LCCN 2022058403 | ISBN 9781032256153 (hbk) |
ISBN 9781032267685 (pbk) | ISBN 9781003289845 (ebk)
Subjects: LCSH: Facility management. | Work environment. |
Quality of work life. | Employees–Psychology.
Classification: LCC TS155 .M528 2023 | DDC 658.2–dc23/eng/20230130
LC record available at https://lccn.loc.gov/2022058403

ISBN: 978-1-032-25615-3 (hbk)
ISBN: 978-1-032-26768-5 (pbk)
ISBN: 978-1-003-28984-5 (ebk)

DOI: 10.1201/9781003289845

Typeset in Times New Roman
by KnowledgeWorks Global Ltd.

Dedication to Lotte Holck

Contents

Foreword

Rianne Appel-Meulenbroek[1] and Vitalija Danivska[2]
[1] *Eindhoven University of Technology, Netherlands*
[2] *Breda University of Applied Sciences, Netherlands*

This is the third book in the book series titled "Transdisciplinary Workplace Research & Management". With this series, we as series editors aim to bundle important insights from the many different disciplinary fields that are studying work
environments and the workers using them. Because so many disciplines are involved in this relatively young scientific research field, the series is focused on
spreading theoretical knowledge, methodological approaches, and developments
amongst workplace scientists and practitioners. Everybody uses some type of
(physical and/or virtual) place to do their work-related activities, and increasingly
people work at multiple locations and spaces throughout the week. They collaborate
within ever-changing teams, projects, environments, and cultural contexts. There
are still many gaps in workplace research that should be filled to provide further
insights into managing different types of workplaces more strategically and how to
align workspaces better to individual employee needs and preferences. These gaps
often exist because they lie at the crossroads of disciplinary expertise and therefore seem (too) hard to grasp. This series hopes to stimulate the take-up of these
research challenges by transdisciplinary teams of researchers and practitioners.

Luckily, transdisciplinary workplace management has seen tremendous growth
in interest both within the research community and industry in past years. The
recent pandemic has raised attention from national governments and local policy
makers to understand how workplaces are more optimally designed in an evidence-
based manner. Thus, it is as vital as ever to increase and strengthen workplace research in academia and practice. This third book in the series provides many ways
to do so. It explains different methodologies that can be used to learn more about
workplaces by both scientists and workplace managers in practice and thus helps
to solve existing gaps in knowledge.

A scoping review on workplace research (Appel-Meulenbroek et al., 2018) showed that the researchers within a certain discipline tend to use only a few specific types of methodologies to gather new data. The transdisciplinary approach of this third book informs and crosses all these disciplines by explaining the relevance of different quantitative and qualitative methods used in many fields and providing examples of applications to help others apply the method correctly. This is of course very helpful for PhD researchers to shop for the most optimal available way to gather data. But it can also bring value to more established researchers by further expanding their methodological options for developing the workplace management field. In addition, each chapter describes why and how this method is relevant to apply in practice by workplace, HRM, ergonomics, facilities, and other types of managers/consultants.

The first two books proved to be a success among researchers and practitioners. And we are thus delighted to add this third book to our series and to exchange knowledge about research methods. We thank the editors and authors of the third book for all their work and wish its readers many insights for future studies and workplace projects.

Reference

Appel-Meulenbroek, R., Clippard, M., & Pfnür, A. (2018). The effectiveness of physical office environments for employee outcomes: An interdisciplinary perspective of research efforts. *Journal of Corporate Real Estate, 20*(1), 56–80. https://doi.org/10.1108/JCRE-04-2017-0012

Preface

Expanding the understanding of methodological
approaches for workplace research and
management in the new era

Ying Hua[1], Chiara Tagliaro[2] and Marko Orel[3]

[1]*Cornell University, USA*
[2]*Politecnico di Milano, Italy*
[3]*Prague University of Economics and Business, Czech Republic*

Multiple forces are shaping work and workplace in a complex way, including eco-nomic, demographic, social, environmental, technological, and more. In the past couple of decades, we have been observing an accelerating evolution of work-places, and an expansion of the scope of workplace studies shedding light on a deeper intertwinement of people, space, and technology. In terms of the physical boundaries, the workplace is no longer a phrase equivalent to office buildings or corporate campuses, but where knowledge work takes place – potentially any place thanks to the high level of connectivity enabled by technology – including both tra-ditional and nontraditional settings. The influence of the sharing economy spawned new workplace settings and business models and, more importantly, pushed the view angle of looking at the workplace beyond its function of accommodating em-ployees and their work activities but rather as platforms and resources for desired outcomes. This shift of view angle unlocked several links to various aspects of peo-ple, teams, and organisations. It made clear the importance of an in-depth under-standing of people and organisations for workplace studies. Meanwhile, companies continue to pursue both competitiveness and resilience in an increasingly uncertain and volatile business environment. They must diligently track a more extensive range of outcomes, including employee health and well-being and various social values. All these new phenomena are calling for new research inquiries and robust methods for collecting evidence and furthering theory development.

While all the above is taking place, the society encountered an enormous challenge from the pandemic. The almost overnight changes to the workplace arrangements started as a temporary solution to keep businesses going under strict public health constraints but turned out to have a lasting effect on companies' understanding of the workplace and their workplace strategies for the future. A hybrid work mode at a certain level, referred to by many as the new normal, became common practice and desired by both employers and employees in many business sectors and globally – the trend towards an increasing level of location autonomy had already started before the pandemic, but was more evident and accelerated after a crisis-induced societal-level "experiment". At the same time, technological advancement is quickly equipping virtual settings as a solid element of work, with still enormous space for innovation and development. In front of workplace researchers is a highly volatile, complex, and exciting era where various aspects of the context for existing workplace theories have evolved. There is an unprecedented level of interaction among different workplace disciplines, e.g., planning and design, business management, human resource (HR), operation, and information technology (IT). They are calling for research with robust methodology and informed accumulation of evidence to guide design and management practices. The wide array of scenarios also presented unprecedented opportunities for expanding the boundary and impact of workplace studies.

Furthermore, another important observation is that the modern workplace continues to give more attention to the diversity of needs and expectations, recognising occupants (as individuals, teams, and organisations) as active participants in the space-making processes to create supportive work environments for more. It is an essential feature of the new workplace compared to earlier models emphasising standardisation and efficiency. The shift from focusing heavily on the average levels and the needs of the 'majority' when evaluating outcomes from workplace interventions creates better opportunities to support diversity, equity, and inclusiveness. Such a shift also calls for learning from multiple disciplines to enrich and expand workplace researchers' capability to study and understand the perception, experience, and behaviour in the workplace both of individuals and groups, to ensure those insights are scientifically sound and robust.

Composing this book is an effort the three of us felt excited to initiate in response to the challenges and opportunities for workplace research and management outlined above, in addition to our individual research. The inception of this book started during the summer of 2021 when we got to know each others' work in more depth and were intrigued by the respective approaches to workplace. The first two books of this Transdisciplinary Workplace Research and Management series had just come out, including one chapter authored by Chiara Tagliaro and Ying Hua. We shared the enthusiasm for a noticeable advancement in the consolidation of workplace research and management. A common feeling resonated among us that, in our different educational paths and backgrounds, extensive exposure was still missing to a broad array of methodologies supporting the field. Therefore, we joined our complementary forces in the attempt to collate a variety of methods that could (i) support both scholars and practitioners in their work; (ii) show the extent to which workplace research and management spans across an assortment of disciplines; and (iii) foster collaboration across fields as well as between research and practice. Our goal is that this volume fosters cooperation already by making scholars and practitioners from different fields understand each other better, first and foremost, by becoming more aware of and familiar with the others' approaches to tackling workplace-related complex issues.

The book starts with a general commentary reflecting this endeavour's purpose and rationale, and offers guidance to navigate the contents. The selection of methods that follows reflects our vision of the next phase of workplace research and its impact, our understanding of the knowledge and methodological gaps in current studies, and a balanced weight of both people and place. A short compendium complements the roundup of methods with useful lessons on how to find an orientation in the method selection process. Finally, some key methodological takeaways will accompany the reader toward the end of this fascinating journey.

In short, we hope to support a transdisciplinary collective effort by our research community to continue developing workplace theories and to support workplace planning, design, and management in the exciting new era.

1 Outlook

Collecting methods for transdisciplinary workplace research and management

Chiara Tagliaro

Politecnico di Milano, Italy

Workplace management is a "collaborative task towards aligning the workplace with the organization and the employees using it" (Danivska & Appel-Meulenbroek, 2022, p. 2). Although it has not been recognized as an autonomous discipline independent of corporate real estate (CRE), facility management, or human resource management, workplace management is progressively gaining a distinct identity. The work of a few scholars that bundled up theories as a foundation of the field (see the two precedent volumes of the book series that hosts this handbook) has been contributing to making it clear that a variety of approaches converge from distinct disciplines into one recognizable set of tools, reasoning, and logic. This isolates workplace management as a specific effort aimed at understanding and improving the way organizations, buildings, and people interact in a holistic and transdisciplinary way.

This happens at a moment when workplaces have been seriously challenged in their very nature, as hinted in the Preface. *What* is the office? *Where* are we working? *When* are we working, and *when* are we *not*? All these questions have occupied the minds of most of the population employed in knowledge-intensive jobs in the past years, and more urgently after the outbreak of the COVID-19 pandemic. Starting from those 'ontological' questions, management of the (physical) work environment is gaining more and more attention in both the industry and academia. Still, the bulk of knowledge on this topic is rather scattered and will benefit from progressive systematization. On the one hand, workplace managers require assistance applying workplace theories in practice and making decisions about the workplace with appropriate information. On the other hand, workplace researchers need to combine all the angles from which workplaces are studied and take advantage of a reference collection of methodologies from different disciplinary areas that are applicable in this context. The multiplication of available data further complicates the matter. Novel opportunities are open to triangulate information from various sources and produce original insights. However, guidance is lacking to support the full exploitation of data through both traditional and innovative methods.

DOI: 10.1201/9781003289845-1

1.1   Foundation axioms and scope

Workplace management shares with other relatively young disciplines a "rich diversity of foundation axioms" (Linstone & Turoff, 2002, p. 15). Methodologies of different names come up frequently, borrowed from "this discipline or that previous study". Often, the choice of a particular method may appear rather casual or occasional. Indeed, as recalled in the Foreword, researchers tend to use only a limited number of specific methodologies (Appel-Meulenbroek et al., 2018). This might be due to the variety of axioms and available methods, which makes it difficult for an inquirer to orient him-/herself in the quarrel about what technique is the best to gather data and address new problems. Not only are we generally unaware of the wide range of different approaches that may underlie our research questions but we also tend to have limited sight of the potential range of our choice. Background studies, training, experience, and philosophical references any person holds affect the choice.

The following edited handbook has been composed to elevate awareness over methodological approaches that can be used in workplace research and management. Given the variety of disciplines relevant to the said field, the compiled methodological approaches might inspire novel and not habitually used data-gathering approaches in subjects closely interlaced with the contemporary workplace. The attempt is to generate a critical examination of alternative and/ or complementary methods, with emerging differences and conflicts between one another, and to encourage conscious evaluation of the approach to adopt. No one method can fulfil all the requirements for a specific research question, and no one mode is *best* for all circumstances, making it even more important to choose consciously.

1.2   Rationale for the book

This book emphasizes not much the results of a particular application but the reasons why a specific method was used in a certain context and how it was implemented. The idea is that, from this approach, the reader will be able to transpose the considerations to their problems and decide on the applicability of the method to its area of endeavour. Each chapter delineates the inquiring system that has been used to address some issues in the workplace realm. The collection of methods happened quite spontaneously and, as such, we believe the book could be read in different ways, switching back and forth through the chapters, based on each reader's interest. This Handbook, therefore, includes foundational knowledge of different methodological research approaches; innovative evolutions of these methodologies; and their added value for application in various workplace contexts. The volume proposes a hands-on approach and guides the reader throughout the research process until the interpretation of outcomes. The necessary evolution of the office should follow evidence-based decisions on the abovementioned matters, which are only possible through rigorous investigations. This volume supports these investigations, which call for innovative applications of qualitative and quantitative methodologies. Doing so will encourage more solid practices and thorough research agendas in workplace design, management, and use.

This introductory chapter develops as follows. First, it discusses *evidence-based* approaches starting with a discussion on using data to back meaningful decisions for research and practice. Afterwards, it gives a panoramic view of how data collection and elaboration methods evolved and can now support workplace strategies and decisions. Then it briefly touches upon the clash between quantitative and qualitative methods by suggesting that the workplace realm might become an exciting ground for a more 'peaceful' exchange between the two approaches. The second part of the introduction anticipates the methods herewith collated. The emergence of *multiple perspectives* is commented on in the following section, which gives an overview of the various interpretation lenses that can be taken while exploring the book and will help the reader find their own mental order across the different contributions.

The whole volume is embedded in the contemporary world of work. As such it encompasses a variety of methods that can capture the most recent transformations undergoing in the workplace. It is worth noting that, when buildings are constructed for particular purposes and at a certain time in history, they tend to have many similarities or, at least, they apply similar design strategies (Becker, 1990); hence, it is likely that they share a similar view towards the future and a similar predisposition to analysis. Even though this book does not claim completeness, the selection of methods herewith included will help address many of the common issues in today's workplace realm, which are summarized under a few *themes and threads*. Finally, the *set-up* of the single chapters is described, and the diversity of the contributors is shown to reflect the richness and complexity of research and management in the workplace.

1.3 Evidence-based approaches

Buildings are socially negotiated solutions (Watson, 2003) designed to meet the typical needs of the human being for shelter and to support the development of their activities. To manage these negotiated solutions, a holistic approach is fundamental as it integrates objective-quantitative data with stakeholders' subjective opinions, preferences, and experiences (for an in-depth discussion about the term 'stakeholders' and synonyms see Danivska & Appel-Meulenbroek, 2022, p. 8). This requires periodically capturing the "quality of use of the built space" (Foti, 2005, p. 145), that is the quality of the relationship between buildings and people.

Poor quality and misuse of the built space will negatively impact the result expected by both occupants and operators in the built environment, namely finding the appropriate support for their activities in the space. On the contrary, the key to successful results corresponds to reasonable levels of information and communication between the parties, which is possible only when decisions and strategies are grounded on evidence-based information. Consequently, building users benefit from gathering *data* through specific methods and approaches to align the built environment conditions with their respective interests.

1.3.1   The use of data to back decisions

Ackoff (1989, p. 3) defines data as "symbols that represent the properties of objects, events, and their environments". Making sense of data has been a constitutional act for human beings since the earliest societies. Data measurement and analysis are embedded in a process encompassing the collection of records from various sources and subsequent interpretation to unveil information that is useful to make decisions. Data elaboration has always been at the basis of human behaviour in its attempt to get to know the surrounding world and gain knowledge and understanding. For much of history, mankind's highest achievements arose from conquering the world by measuring it (Mayer-Schönberger & Cukier, 2013). The first traces of data storage can be recognized in the prehistoric ages when tribespeople used sticks and bones to track records of trading activities or food supply (Marr, 2015). The Library of Alexandria might be considered the most extensive data collection in the ancient world.

The beginning of data analysis dates back to the first statistical purposes in ancient Egypt when periodic censuses were necessary to manage building pyramids (Mayer-Schönberger & Cukier, 2013). Retrieving data through censuses has always been used worldwide, especially for governmental planning activities, such as taxation. Mayer-Schönberger and Cukier (2013) report the example of analysing population growth by the district for governments to determine the number of necessary hospital facilities in different areas. Therefore, census takers are already used to collect and process data to increase its usefulness by transforming it into information.

Over time, measurement techniques have been refined, and data have been multiplying. The advent of computers disrupted traditional ways of dealing with data by making gathering a lot easier and less time- and resource-consuming. Meanwhile, this new computational capacity has also revolutionized our approach to data analysis, by opening up new evidence of correlations between phenomena (Mayer-Schönberger & Cukier, 2013). Since the advent of computers, the evolution has gone through business intelligence, data centres, and the internet, until the emergence of Big Data (Marr, 2015). Nevertheless, all data need wise elaboration in order to provide useful insights, explain events, and support decisions. Automated data-elaboration systems cannot generate the understanding and wisdom necessary for learning and development. Ackoff (1989, p. 3) while drawing his renowned Data-Information-Knowledge-Wisdom pyramid states that: "Information [...] answers to questions that begin with such words as *who, what, why, when, where, and how many*". Instead, knowledge answers how-to questions while providing *instructions* to control different systems. Understanding answers to *why* questions thus enables the detection of errors, the identification of their causes, and, eventually, their correction. Finally, wisdom is the ability to employ judgement and evaluate a choice concerning the amount of progress it will make possible. Going from data to wisdom necessitates the adoption of sound methods.

1.3.2 Data collection and elaboration methods in the working environment

Many companies nowadays use data-driven decision-making for their core business operations – e.g. Amazon and Netflix, among others (Miller et al., 2014; Waber, 2013). Curiously enough, oftentimes they forget to apply data-driven decisions inwards, i.e. to face issues regarding the way people work, and the management of their workplace, even though a variety of data might theoretically be available. Data analysis has been happening in the working environment for many years, at least since the middle of the twentieth century when managers started getting appointed to organize and control corporate physical spaces (Danivska & Appel-Meulenbroek, 2022). The earliest methods encompassed data from human observation (Waber, 2013). Subsequently, we augmented data collection tools by applying interviews and surveys, which better encapsulate the people's components of the workplace. Now plenty of new data about the workplace are available from various sources (Waber, 2013), including psycho-physiological measurements, real-time sensor data, and more.

The extension of data sources makes data collection and elaboration further intricate and complex. If the availability of multiple data sources allows relief from biases typical of social science disciplines (Mayer-Schönberger & Cukier, 2013), now the issue lies in how to merge all the different data sources in a meaningful way and obtain a sound understanding of data; there is a strong need "to balance between personal judgment and datafied assessment" (see Chapter 4 of this book). Moreover, "wisdom-generating systems are ones that men will never be able to assign to automata" (Ackoff, 1989, p. 9), meaning that human action is fundamental in the process and it cannot prescind from thorough awareness of and capacity to combine the available methods.

1.3.3 About the complementarity between qualitative and quantitative methods

In very general terms, the inquiry process happens throughout a 'method' or several consequential stages. First, an event or a raw dataset presents itself in the real world (Linstone & Turoff, 2002). Originally from the Latin 'given', *data* represent what is available to the senses (Uher, 2019). Nevertheless, this does not mean that they are ready to be collected. Each individual will be driven to recognize only specific 'objects' as relevant data, based on his/her own subjective perceptions, prior experiences, and field of investigation. Therefore, data generation and collection are an act of choice that must be carefully undertaken in every research setting (read more about how to choose a method in Chapter 14).

Second, some transformation or filtering of this dataset is applied to make it compatible with input into a 'model'. The model corresponds to some set of rules or a structured process (e.g. an algorithm, heuristic principle, or theoretical framework) that is functional to elaborate the 'input data' and convert it into 'output information'.

An encoding process must occur based on agreed schemes (Uher, 2019). Namely, this phase consists of transforming, reorganizing, and interpolating the data by putting it in a way (i.e. the 'model') that carries some new meaning or adds some levels of understanding compared to the previous stage of data generation and collection. Finally, the information obtained herewith will need another filter or transformation round to be presented in a form that is useful for making decisions or recommendations (e.g. for policymakers, designers, company managers, etc.).

Throughout this process, the debate around quantitative and qualitative methods usually comes up more or less implicitly. The discussion revolves around whether data should be explored utilizing quantification through measurement instruments or require the recognition of specific qualities by human interpretation. The two approaches are normally considered in contrast with one another. The first is typically associated with a *natural* conception of causation (*N-causation*); therefore, it implies the account of a systematic pattern of behaviours or happenings, on the model of experimental approaches adopted in the natural sciences (Howe, 2011). The latter, instead, is typically associated with an *intentional* conception of causation (*I-causation*), meaning the account of norm-governed institutions and practices behind behaviours or happenings, on the model of interpretative approaches applied in the social sciences (Howe, 2011).

Often, qualitative methods are attributed only an auxiliary role in exploring new phenomena and/or descriptions. In contrast, quantitative methods have the more 'prestigious' function of understanding and explanation. To subvert this view, the concept of "mixed-methods interpretivism" has been proposed (Howe, 2011), according to which experimental-quantitative methods play the basic role of recognizing patterns. Besides, qualitative-interpretive methods are responsible for reaching a deeper understanding of causation. This would show how quantitative methods can be used to investigate I-causation, while on the flip side, qualitative methods can be employed to disentangle N-causation. These contrasting positions pose doubt about what 'scientific' research is and whether non-scientific research may even exist.

The boundaries between research and evidence-based practice are sometimes blurred in the workplace realm. Applied research based on hard measurements intertwines with overarching psychological and philosophical theories on people's interactions with each other and with the space. In particular, what is unique about workplace investigations is that they often require a combination of the two approaches, which, in the end, come out not as an alternative to one another but rather as complementary. The combination of quantitative and qualitative methods is already frequent in workplace research and management and deserves further exploration and exploitation. This volume takes the first step in this direction by presenting multiple perspectives to handle workplace-related data.

1.4   Multiple perspectives

A book needs to file the chapters that compose it in some order. However, deciding the most interesting, reasonable, and/or natural way of ordering the various contributions in this specific volume was challenging. Also, the collection of methods is far from comprehensive and exhaustive as it was not subject to particular criteria other than "what happens to be useful and original to address contemporary workplace matters". Therefore, we found it stimulating to highlight multiple perspectives through which one can create a personal path throughout the reading experience.

These perspectives may also offer hints on practical and more theoretical ways to approach the methods. Below they are described and grouped by those that may suggest when, where, and how to apply a method (Table 1.1); and those that will tell instead who and why to use a method (Table 1.2). The first set of perspectives gives information about data collection and processing, whereas the second set collects the origins of the different methods and outlines the logic for their choice.

With these instructions, we also intend to provide a ready reference to support multi- and mixed-method approaches and triangulation of methods.

1.4.1 When, where, and how to apply a method

The order chosen by the editors aims to reflect a *spectrum between the human component and the spatial component of the workplace.* The first chapters focus primarily on people to show that, first and foremost, researchers and practitioners should acknowledge who is using the space and with what purpose. The researcher's own human presence is recognized within the inquiring process as an essential element that may affect the investigation results and even become the same investigation object. The last chapters, instead, focus more on the physical elements within which people behave while working. Of course, all the methods, in the end, aim at disentangling the difficult relationship between people and space during work.

However, the reader will find a logical flow from methods that primarily look at people's feelings and perceptions (e.g. ethnographic methods) to those that analyse the spatial dimension (e.g. Space syntax). This flow is curiously reflected also in the writing style; an attentive reader will appreciate that the first chapters are written in first person, whereas the voice becomes more impersonal and objective towards the end of the book, where the authors take the third person to describe how they approached and applied the different methods. By going through the chapters in this order, the reader might as well notice a progressive change in tone from a more theoretical to a more practical attitude. A gradient appears from the ethnographic field, which is interesting on a scientific level and offers operational instruments to a practitioner as a secondary outcome, to the spatial approaches described towards the end, which are more directly transferable in practice and reveal scientific findings as a subsidiary impact.

Beyond this general note, a few other points of interest are worth mentioning. First, is the emergence of different *granularity, or scale, in the data collection process*, despite generalization always being a scientific goal. Some methods gather data about individuals, namely one's preferences and emotions (e.g. Stated Choice Experiments; Autoethnography), whereas other methods tend to privilege the interactive dimension between two subjects (e.g. Social Network Analysis). Some

Table 1.1 Practical information about data collection

Chapter/method	Granularity/scale of data collection	Type of data	Kind of evidence/data elaboration	Timestamp of data
2. Workplace autoethnography	Person	Qualitative	Narrative of an individual experience (either single or co-authored)	Present
3. Affective ethnography	Person/ organization	Qualitative Description of an experience through participant observation + Quantitative Organizational data like statistical records, salary structures, organigrams, company documents on values, policies and strategies, employee satisfaction reports, etc.	Autobiographical accounts of and interviews with the participating employees, work diaries, and participant observation in extensive fieldwork notes	Present
4. Digital ethnography	Person	Qualitative (ethnographic observations) + quantitative (computational network analysis)	Fieldnotes during participant observation and technical walkthroughs, interview transcripts + network graphs	Present
5. Critical discourse analysis	Person/ organization	Qualitative and quantitative (spoken, written, or otherwise depicted forms of texts)	Theoretical sampling (also assisted by computerized operations)	Present or past
6. Diary studies	Person	Predominantly quantitative self-reports over an extended period of time (e.g. questionnaires, tally sheets, physiological measurements, or pictorial scales)	Quantitative data analysis (statistical) – recommended consultations with someone profoundly experienced	Present (longitudinal studies), ideally, do not contain any retrospective assessment of certain periods, but the assessment aims at the current experience

(Continued)

Table 1.1 (Continued)

Chapter/method	Granularity/scale of data collection	Type of data	Kind of evidence/data elaboration	Timestamp of data
7. Cluster analysis	Person/group	Quantitative – survey data (including self-reported data)	Group observations through hierarchical and non-hierarchical algorithms to maximize within-cluster homogeneity and between-cluster heterogeneity	Present, past, and future
8. Stated choice experiments	Person	Quantitative –revealed (real life) or stated (controlled hypothetical situations) choices/preferences	Statistical models (e.g. multinomial logit, latent class model, mixed logit)	Future Punctual data (data are acquired at one moment in time)
9. Delphi method	Person/group	Qualitative or semi-quantitative	Qualitative or semi-quantitative analysis	Future
10. Social network analysis	Relationships/interactions between members of a network	Quantitative relational data from questionnaires, interviews, observations, and artefacts through purposive sampling (necessary to have information from every actor in the network)	Network graph visualization Quadratic assignment procedure and multiple-regression Quadratic Assignment Procedure analysis	Present and past
11. Surveys	Person and space	Quantitative	Statistical analysis	Present
12. Space Syntax	Space	Parameters of spatial configuration that both describe a space (central, isolated, large, elongated, open/close) allow for comparison to human behaviour (location, activity, perception)	Statistical analysis	Present
13. Journey Mapping	Person and space	Semi-structured user interviews, user testing, facilitated co-creation workshops and secondary data from desk research	Journey map: a schematic visual document, typically matrix-like, used as a proxy to represent one or more user experiences within space	Present or future

Table 1.2 Origin of the methods and logic for their choice

Chapter/method	Original disciplines	Underpinning theories/ philosophical approaches	Output (goal)	Outcome (helpful for)
2. Autoethnography	Ethnography, autobiography	Positioning theory	Reveal individuals' perception towards physical space and its change due to specific factors	Understand a particular cultural occurrence
3. Affective ethnography	Ethnography and cultural anthropology	Performativity Embodied phenomenology (Merleau-Ponty)	Go native and take the perspective of the participants usually with reformative ambitions	Understand the complexity related to diversity
4. Digital ethnography	Human geography, sociology, anthropology, nursing, educational sciences, and, more recently, business, and management studies	Emic perspective Grounded theory	Learn how people do things Elicit subjective meanings	Interpret the penetration of digital platforms into the workflows and taskscapes of employees
5. Critical discourse analysis	Linguistics	Social constructivism	Shed light on the practices of subjectivity in creating concepts (e.g. physical, remote, and hybrid work)	Demystify ideologies and help the dominated people towards emancipation from the associated problems
6. Diary studies	Psychology, anthropology, history, and literature	Within-subject vs between-subject approach	Unveil the fluctuating nature of thoughts, feelings, and behaviours, their antecedents, and their dependence upon situational conditions	Increase accuracy in the formulation and testing of hypotheses, the assessment of phenomena (thoughts, feelings, behaviours) that are dynamic, allows to test and reject causal explanations
7. Cluster analysis	Marketing, biological sciences and genomics, operation management	Post-positivist approach	Group workers (i.e. profiling) based on multiple dimensions	Create workers' profiles based on their relations with the workplace

(Continued)

Table 1.2 (Continued)

Chapter/method	Original disciplines	Underpinning theories/ philosophical approaches	Output (goal)	Outcome (helpful for)
8. Stated choice experiments	Marketing, health, transportation, and tourism	Information integration theory, probabilistic choice theory, Random utility theory	Describe, explain, and predict the choices people make between two or more discrete alternatives based on their individual preferences	Identify the relative importance of the attributes of the choice alternatives, predict the probability that certain alternatives will be chosen
9. Delphi method	Defence, technology, healthcare, and medical applications	Combination of positivist and constructivist approach General Theory of Consistency (GTC)	Reach stability in opinions and ultimately facilitate the convergence of different viewpoints	Long-term forecasting; gathering current and historical data; exploring planning options; putting together the structure of a model; delineating pros and cons of policies; uncovering causal relationships; exposing priorities of personal values
10. Social network analysis	Organization studies	Least effort theory and homophily theory	Map patterns of relationships among interacting members	Indicate how an individual is connected to others, also show the cohesion of a network
11. Surveys	Social science research	Person-environment fit theory	Create a snapshot of the occupant perception in the workplace	Describe the basic characteristics of experiences of a population
12. Space syntax	Architecture, urban planning, archaeology, neuroscience, and biology	Space syntax	Treat space as a set of parts (rooms, streets) interconnected through permeability or visibility (doors, windows, and junctions), forming spatial networks	Understanding the relationship between workspace and human behaviour, and derive suggestions when designing space
13. Journey mapping	Business management, user experience (UX) design, and service design	Human-centred design actor-network theory	Describing an experience (including user needs and challenges) from the point of view of the user	Interpreting the reasons why people behave in a certain way in space

other methods look at the group or organizational dimension (e.g. Delphi; Critical Discourse Analysis) and aim to aggregate data beyond the individual data point to recognize overarching trends (e.g. Cluster Analysis).

One additional element to point out is the *type of data* and *kind of evidence* that is produced through certain elaboration techniques. This goes beyond the schematic distinction and battle royal between qualitative and quantitative techniques. As anticipated above, most methods combine both qualitative and quantitative data and elaborate them in such a way that the researcher can interpret them, sometimes thanks to the aid of computerized systems. Intriguingly, many methods require the interaction of different actors (either additional researchers or between researcher and observed individuals) into an interactive process. This reminds us that evidence is never apparent but requires repeated attempts of interpretation from multiple viewpoints to be deeply understood.

One final point regards the focus over time or the *timestamp of data*. Some techniques mostly use data that look at the present, almost simultaneously or very close to the time of analysis and interpretation (e.g. Space syntax). On the contrary, other methods utilize data created in the past, even in times very far from the moment they are elaborated (e.g. Critical Discourse Analysis). Finally, specific methods can be used to simulate future scenarios and generate 'what-ifs' situations projected into the future (e.g. Stated choice experiments; Delphi method).

1.4.2 Who and why to apply a method

Different methods' attention on people or space reflects the disciplines where the method has been incepted. Indeed, the methods investigating primarily people and their behaviour originated from disciplines such as ethnography, anthropology, biography and literature, linguistics, and psychology, among others. Conversely, methods addressing first the spatial features of the workplace are related to disciplines like architecture, urban planning, service design, neuroscience, and technology.

Partially depending on the *disciplinary affiliation*, most of the methods are rooted in *underpinning theories or philosophical approaches* that explain the way these methods are conceived and applied, their reliance on quantitative or qualitative data, and the way they require these data to be elaborated. Most of them are embedded in a constructivist approach. However, it is recommended to look at these theories while consulting the previous books of this series to find connections between theories that mainly belong to disciplinary areas other than the workplace and those that have their origins within the workplace domain.

Finally, we can distinguish between the *direct output* of a method and its *outcome*. By output, we mean the immediate goal one can reach when applying one method, for instance, understanding individuals' perceptions and their change given different conditions, explaining and predicting choices, etc. By outcome, we intend instead the indirect effect that one may have the ambition to reach when adopting a method (i.e. what is the method useful for?), namely the kind of *wisdom* that the method enables. For example, inform future developments based on values, identify the relative importance of alternatives, test causal explanations, and so on.

1.5 Themes and threads

Throughout the pages, the book discloses many of the most relevant contemporary themes in workplace studies and practice, which recur in more than one chapters. Especially, the volume focuses on the most pressing questions regarding the relationship between the spatial component of the workplace, including its progressive hybridization with other physical and virtual places, and its users, being public organizations, private companies or start-up businesses, and solopreneurs. These questions do not have an impact only in the research realm but translate into very concrete practical matters and deserve to be tackled through careful analyses.

It is recommended to start from the first chapter on 'Autoethnography' (Chapter 2) to set the stage. Here the reader will find a broad story of how work has been transforming lately, especially in the wake of the COVID-19 pandemic. *COVID-19* has indeed played a crucial role by impacting how people work, and making them interact much more with screens than with people. The chapter on "Digital ethnography" (Chapter 4) explains well how we cannot prescind nowadays from investigating this aspect and shows that new research methods should be integrated with traditional ones. Consider, for instance, that technical walkthroughs are now available besides physical walkthroughs. In an era where digital and analogue ways of interacting are deeply intertwined, it makes sense to adopt mixed-method approaches that can illuminate the multiple facets of digital labour.

The *dematerialization and digitalization of work* exacerbated by the hit of the pandemic is addressed in different chapters, including "Critical discourse analysis" (Chapter 5), as something partially independent from physical space but still happening in some other (virtual) dimension that needs to be coherent with the company culture, ethics and power relations, and its approach to space. This will also affect the way people interact with one another and their social ties, as the chapter on "Social network analysis" underlines (Chapter 10).

Multiple chapters highlight the need for understanding spatial and personal relations *within and beyond the traditional boundaries of 'the office'*. The chapter on "Cluster analysis" (Chapter 7) shows how different work locations might determine alternative work patterns and, therefore, foster multiple clusters of workers who use work locations for different purposes. The chapter on "Workplace autoethnography" (Chapter 2) points out the existence of new working spaces (e.g. coworking spaces) and homes in the balance across multiple workplaces. In addition, "Social network analysis" (Chapter 10), "Space syntax" (Chapter 12), and "Journey mapping" (Chapter 13) all bring attention to the use that people make of different workplaces, highlighting how research and practice need to understand the human-space relation better, especially considering how space can affect people's behaviour beyond traditional offices.

The increasing relevance of ethnographic research is evident given the attention to the aspects of *diversity and inclusion* that are becoming more widely embedded in the workplace realm. The varied human component of work is a crucial focus of all the chapters; however, the chapter on "Affective ethnography" (Chapter 3) places particular attention on the role of the researcher in understanding the dynamics of the contemporary workplace. Besides the well-known Hawthorne effect, which explains how the presence of researchers can affect the behaviour of the observed population, this chapter introduces the underestimated factor of how the same researcher can feel within the research process – especially during some immersive experiences like that of covered participant observation.

The researcher's feelings might indeed raise important dimensions in research including sensitivity to data generation and the bodily sensation of the working body. The same reflections could also be drawn for what concerns the role of professional workplace operators (facility managers, CRE managers, human resources managers, architects, and designers). Problems potentially rising from the growing diversity of the working population, such as racism and ethnic prejudice, can be found also in the chapter dedicated to "Critical discourse analysis" (Chapter 5).

This growing diversity enhances the importance of *individuals* and their respective preferences and needs. The chapter on "Stated choice experiments" (Chapter 8) addresses the idea that each person carries their own desires and may have distinct approaches and preferences about how to work. This implies additional attention to what concerns the *health and well-being* of people, as the chapter on "Diary studies" underlines (Chapter 6).

The fundamental importance of measuring workplaces 'performance' and their fit with people's characteristics and needs by integrating various measures and indicators is mentioned in chapters about "Surveys" (Chapter 11) and the "Delphi method" (Chapter 9). They consider the present and the past's role while discussing the alignment of the workplace with the organization and the employees using it and how workplace management helps reach this goal.

1.6   Set-up and authors

To respect the scope of a handbook, which is that of being a tool assisting the reader in some learning process, and to offer a reference work, all the chapters follow a similar structure, even though some flexibility was given to the authors to personalize their style and make it coherent with the method itself.

On the whole, each chapter develops into five subsequent sections. This first section introduces the *background* of the method. It gives context on how the method can support (a) advancements in the workplace research field; and (b) evidence-based decision-making in practice. This section introduces where the method is coming from (who are the people that initiated it), in which fields it has been used, its disciplinary roots, evolution over time, etc. – plus the essential tools for its application. This section also describes the basic assumptions of the method and the general approach to its application. In the second section, the reader finds *arguments* on how this method has been and/or can be applied to workplace research and/or management. Here the authors explain why it makes sense to implement this method to solve a workplace-specific question/issue. Especially, one can read here specific comments on how this method may help balance out the digital and physical work environment and help face the most recent challenges related to physical vs. remote vs. hybrid types of work. Moreover, the section discusses what competencies are required when applying the method, how long it may take to apply, etc. The third section hosts *examples of the method's application* in one or more specific workplace cases in research or practice. In most chapters, it reports on the experience (often done by the same author or authors) while applying that method to solve a workplace-specific question/issue. These concrete examples, of how the method has been used already, include details about the objectives of its use, data collection strategies, and analysis. The resulting empirical outputs are shown only briefly because the focus should not be on the outputs and outcomes that have been obtained but rather *on the way* they were obtained, stressing the usefulness and applicability of the method. The fourth section presents *implications for both research and practice*. This section provides a critical discussion on why each method can be favourably applied in workplace studies and illustrates how this would help address certain gaps in workplace research and/or management. On the one hand, ideas are proposed on how *scholars* may benefit from the described method if it is widely applied in academic research. Recommendations are also provided for researchers to apply this method in their studies. On the other hand, suggestions are made on how *practitioners* could benefit from the described method if it is widely applied in the professional field. Brief guidelines for practitioners to transfer the basics of each method in practice are available here. Finally, the chapters end with *conclusions*, including key takeaways for the employment of the method (e.g. what kind of questions/issues it is most suited for, either in research or in practice, or both) and some pros and cons to its use. Possible limitations and risks, when present, are highlighted in the described method, and methodological suitability for workplace research is discussed as a final remark. Most chapters are complemented, besides the bibliography, by a short list of *further reading* sources recommended by the authors based on their expertise. These additional references can incorporate both the own research of the authors and other research relevant to situate the method in broader inquiry areas. This is intended to act as an initial expansion of the literature for those who want to learn more about a specific method, before adopting it themselves.

The book ends with *two final chapters* with a different outline than the previous ones. The first of the two proposes an operative guideline for young researchers, scholars new to the field, or practitioners unfamiliar with the evidence-based inquiry approach. The concluding chapter wraps up the main concepts that emerged throughout the book and reflects upon the potential impact of this contribution to both literature and practice.

We are proud to have collected chapters from authors that hold different positions in various educational institutions worldwide. Table 1.3 lists the authors of all the chapters. We believe their diversity expresses well the range of approaches and the multiple paths workplace management can take.

Table 1.3 An overview of the authors

Chapter	Authors	Country	University/organization
Foreword	*Appel-Meulenbroek, R.*	Netherlands	Eindhoven University of Technology
	Danivska, V.		Breda University of Applied Sciences
Preface	*Hua, Y.*	USA	Cornell University
	Tagliaro, C.	Italy	Politecnico di Milano
	Orel, M.	Czech Republic	Prague University of Economics and Business
Outlook	*Tagliaro, C.*	Italy	Politecnico di Milano
Workplace autoethnography	*Orel, M.*	Czech Republic	Prague University of Economics and Business
Affective ethnography	*Holck, L.*	Denmark	Copenhagen Business School
Digital ethnography	*Ritter, C.S.*	Netherlands	Karlstads Universitet
Critical discourse analysis	*Shadnam, M.*	Iran	Sharif University of Technology
Diary studies	*Soucek, R.*	Germany	MSH Medical School Hamburg
	Weber, C.	Switzerland, UK	Zurich University of Applied Sciences and University of Surrey
	Gunkel, J.	Germany	Hochschule Fresenius, University of Applied Sciences
	Degenhardt, B.	Switzerland	University of Zurich
Cluster analysis	*Migliore, A.*	Italy	Politecnico di Milano
	Rossi-Lamastra, C.		
Stated choice experiments	*Appel-Meulenbroek, R.*	Netherlands	Eindhoven University of Technology
	Kemperman, A.		
Delphi method	*Tagliaro, C.*	Italy	Politecnico di Milano
Social network analysis	*Zhou, Y.*	USA	Virginia Tech
Surveys	*Hua, Y.*	USA	Cornell University
Space syntax	*Koutsolampros, P.*	UK	University College London
Journey mapping	*Iadarola, A.*	USA	New York University
Compendium	*Shepley, M.*	USA	Cornell University
Wind-up	*Orel, M.*	Czech Republic	Prague University of Economics and Business

References

Ackoff, R. (1989). From data to wisdom. *Journal of Applied Systems Analysis, 16*, 3–9.

Appel-Meulenbroek, R., Clippard, M., & Pfnür, A. (2018). The effectiveness of physical office environments for employee outcomes: An interdisciplinary perspective of research efforts. *Journal of Corporate Real Estate, 20*(1), 56–80. https://doi.org/10.1108/JCRE-04-2017-0012

Becker, F. (1990). *The total workplace: Facilities management and the elastic organization.* Van Nostrand Reinhold.

Danivska, V., & Appel-Meulenbroek, R. (2022). Collecting theories to obtain an interdisciplinary understanding of workplace management. In V. Danivska & R. Appel-Meulenbroek (Eds.), *A handbook of management theories and models for office environments and services.* Routledge.

Foti, G. (2005). La qualità d'uso dello spazio costruito. In A. Violano (Ed.), *Strumenti e metodi per la gestione della qualità del costruire. La qualità del progetto d'architettura.* Alinea.

Howe, K. R. (2011). Mixed methods, mixed causes? *Qualitative Inquiry, 17*(2), 166–171. https://doi.org/10.1177/1077800410392524

Linstone, H. A., & Turoff, M. (2002). *The Delphi method. Techniques and applications.* https://web.njit.edu/~turoff/pubs/delphibook/delphibook.pdf

Marr, B. (2015, February 25). *A brief history of big data everyone should read.* World Economic Forum. https://www.weforum.org/agenda/2015/02/a-brief-history-of-big-data-everyone-should-read/

Mayer-Schönberger, V., & Cukier, K. (2013). *Big data. A revolution that will transform how we live, work and think.* John Murray.

Miller, R., Casey, M., & Konchar, M. (2014). *Change your space, change your culture. How engaging workspaces lead to transformation and growth.* John Wiley & Sons.

Uher, J. (2019). Data generation methods across the empirical sciences: Differences in the study phenomena's accessibility and the processes of data encoding. *Quality & Quantity, 53*, 221–246. https://doi.org/10.1007/s11135-018-0744-3

Waber, B. (2013). *People analytics. How social sensing technology will transform business and what it tells us about the new world of work.* FT Press.

Watson, C. (2003). Review of building quality using post occupancy evaluation. *PEB exchange. Programme on educational building 2003/03.* OECD Publishing. https://www.oecd-ilibrary.org/docserver/715204518780.pdf?expires=1667813924&id=id&accname=guest&checksum=31EA1183649F7381FE9E6E9AEDDA7B10

2 Workplace autoethnography

Exploring the place through aspects of the self

Marko Orel

Prague University of Economics and Business, Czech Republic

2.1 Background

Autoethnography is a qualitative research approach that comprehensively describes and systematically analyzes the personal experience to understand a particular cultural occurrence (Ellis et al., 2011). Spry (2001) has defined it as a self-narrative that tends to critique the situated self with others in various social contexts and settings. The observations emerge from the autoethnographer's bodily standpoint that continually recognizes and interprets the residue traces of the specific cultural context in which the scholar finds him or herself. Moreover, autoethnographers share a firm belief that the personal experience is infused with a particular set of norms and expectations. They engage in self-reflection to identify the intersections between the self and the social life (Adams et al., 2017). These experiences are commonly narrated in stories that seek to engage readers aesthetically, emotionally, politically, and ethnically (Tillmann, 2009).

Wall (2006, p. 147) has stressed that the importance of autoethnography as a traditional scientific approach "requires researchers to minimize their selves, viewing the self as a contaminant and attempting to transcend and deny it." However, Holman Jones (2016) has defended the role of autoethnography as the approach brings the personal, the concrete, and an emphasis on storytelling as a framework for obtaining the data from the perspective of an individual. While authors like Holt (2003) criticized autoethnography for being introspective, overly individualized, and self-indulgent, Holman Jones (2016) emphasized that no theory is a body of knowledge being a static and somewhat autonomous set of ideas and practices, but instead theorizing being a movement-driven and evolving process that needs to have an ability to link concrete and abstract. In other words, theory and the process of storytelling share an interanimating and somewhat reciprocal relationship, enabling a scholar to explain the nuances of an experience and the happenings in a particular setting or environment.

Autoethnography is grounded in postmodern philosophy and linked to a growing debate about the role of reflexivity in social research (McIlveen, 2008; Wall, 2006). Ellis and Bochner (2003) have justified that the "auto" is the focal point of the autoethnographic study, moving away from autoethnography being viewed as indicative of its introspective narrative (e.g., Denzin, 1999; Hayano, 1979;

DOI: 10.1201/9781003289845-2

Reed-Danahay, 1997), and consequently paving the way for contemporary evocative autoethnographers who center their portrayal narrative on their oneself in a way that contributes an understanding to the greater culture (Doloriert & Sambrook, 2012). While evocative autoethnography has dominated the discipline since the late 2000s, Anderson (2006) proposed a more consistent approach with traditional ethnographic practices. He justified the use of analytic autoethnography where the autoethnographer (i) becomes a full member in the selected setting; (ii) gains visibility as a member in published texts, and (iii) shows commitment toward developing a theoretical understanding of a broader social phenomenon.

2.2 Argument

The beginning of 2020 brought one of the most extensive economic and social upheavals in modern history. While the Covid-19 pandemic can be understood as a global disruptor of virtually everything connected with contemporary society, it changed and transformed everything that we as humanity took for granted. The novel virus shifted and stirred the relationship of individuals with the living environment, from the daily commute in urban or rural areas to how individuals tend to work. The pandemic has compelled organizations to implement telework for their employees to comply with the need for physical distancing (Chong et al., 2020), resulting in what could be described as one of the most extensive work experiments of our time.

While telework has been an integral part of scholarly debates for the last four decades, the spread of Covid-19 brought remote work to a point which could not have been anticipated, either by workers or employers, and from early on this trend has affected workers' occupational health (Bouziri et al., 2020). The first studies on the anticipated health effects resulting from the immediate introduction of telework (e.g., Buomprisco et al., 2021; Dolce et al., 2020; Nagata et al., 2021; Schur et al., 2020) mainly focused on larger samples and studied a group perspective of remote work. However, these studies somewhat neglected to offer a more detailed account of how a selected workplace – a home or a flexible office such as a local coworking space – can transform work, and what an individual's lived experience can tell us about the workplace itself.

That said, the autoethnographic narratives can be a strapping tool for capturing the authentic lived experiences of individuals that find themselves in a selected setting and affected by a particular disruption (Carroll, 2020; King, 2019; Oswald et al., 2020). These experiences can be narrated either by an individual from both first- or third-person points of view, or they can be co-authored by several individuals involved in the same lived experience. While several autoethnography accounts are being used in the context of a workplace where a particular scholar worked, the approach has not yet been fully explored across various disciplines as a suitable tool to understand workplace environments from within. With knowledge work becoming more individualized than before, the autoethnographic accounts present the ability not only to share an individual's living experience, but also the relationship of that individual with the environment where work is being performed within a selected period. As workplace autoethnography requires a rather systematic inquiry approach to data analysis, the analytic approach appears to be the most efficient way to gain a comprehensive insight into a contemporary workplace and workers within.

Therefore, the following chapter explores and demonstrates autoethnography's potential as a methodological framework in the context of a particular work setting by observing both self, lived experiences and personal surroundings. It further argues that autoethnography can be used to understand inner processes within a selected workplace that occur in one's cultural encounters. The chapter first narratively analyzes the foundations of workplace autoethnography by overseeing past studies that focused on a particular subject within a selected workplace setting. In the second part, the chapter presents an autoethnographic account of a study where the author has been seeking an understanding of how a highly flexible and community-based workplace can gradually transform throughout the Covid-19 pandemic, and of how an individual's perception toward the physical space (as well as toward other workplace users) can change due to pandemic-related factors.

2.3   Foundations of workplace autoethnography

Autoethnography can be used to share a light on different, often unseen aspects of workplaces. According to Spry (2001), autoethnographic texts reveal the fractures, sutures, and seams of self-interacting with other participants in a selected social context researched within a particular lived experience. Autoethnographers commonly gaze back and forth at themselves and at the culture to create (and possibly resist) meaning with other individuals or groups of individuals. According to Adams et al. (2017), the latter can be accomplished by (i) understanding the selected setting through stories that enable researchers to be conscious of the surroundings; (ii) understanding the culture that tells researchers who they are (and whom they are not); (iii) understanding positioning in the link of researchers' body's identity markers; and (iv) having others to understand themselves through researchers' stories. That said, recording autoethnographic accounts requires working at the intersection of ethnography and autobiography by observing and participating in a (cultural) experience using self.

A handful of studies look at a particular phenomenon in the context of individuals' workplace and their daily work lives. By that, we mean that they focus on organizational reality by interpreting personal experiences that frequently carry a negative value (Zawadzki & Jensen, 2020). Yet, organizational autoethnography enables the researcher to intervene as a complete member, capturing a rather complex yet befitting way to understand a selected (work) environment (Doloriert & Sambrook, 2012). However, as we will note in the following debate, workplace autoethnography is still somewhat underused as a method of data inquiry, and oftentimes underutilized in the framework of academic institutions.

Miller's (2002) research accounted for one of the first accounts of using autoethnography to find meaning in the workplace context. In her study, she considered how academic professionals are both processors and performers of collective workplace emotions when coping with tragic events. In November 1999, Texas A&M University had experienced a "bonfire" collapse. The researcher observed that while most of the academic personnel had been tasked with maintaining a semblance of normalcy in the workplace throughout this challenging time, most of them experienced severe emotional upheaval. The subsequent autoethnography at a selected academic institution revealed that emotional labor considers the disconnection between felt and performed emotion. The study concluded that the disconnection could become a heavy burden to bear if workers use surface acting to fetch the organizationally sanctioned emotions during the work itself.

Riad (2007) focused her research on understanding the conflicting nature of becoming a mother and reconciling the obligations of motherhood with work-life, concentrating on maintaining meaningful relationships with other individuals within a workplace, and continuing to portray herself as a strong individual that can equally perform given work-related tasks. Her autoethnographic accounts were recorded five years apart and have been subsequently analyzed to understand how a working mother's position can change and possibly transform. Cohen et al. (2009) made a critique of the prevailing metaphor of work-life balance being problematic by an autoethnographic conversation between three scholars. They offered a conceptualization of the relationship between work and other life-related aspects where emotions, an individual's autonomy in the workplace, control, and self-perceived identity become integral features. The autoethnographic approach enabled the three scholars to seek meaning around metaphors of importing, seeping, integrating, and segmenting individuals lives and to understand how these can be affected by autonomy, self-control, and identity by managing the dynamic interplay of structural constraints.

Kempster and Stewart (2010) co-examined the development of situated leadership practice using co-produced autoethnographic self-observation. The researchers' serial narrative identified the various aspects of a situated curriculum involved in developing leadership practice. It overviewed an intricate process-oriented pattern of relational and dialogical engagements within a workplace, which gradually changes over time. Cullen (2011) emphasized how researching workplace spiritualization can challenge the organizational ethnographer, demonstrating the use of autoethnography to craft a representative account of an individual's spirituality within a selected workplace in different contexts, inter-relationships, and perceptions of personal responsibility. Autoethnography as a data-inquiry method has been shown to induce feelings of a researcher's "outsider-ness" in shared workplace scenarios, enabling scholars to understand workplace spirituality from within.

Sobre-Denton (2012) examined systematic discrimination and workplace bullying with the use of autoethnography as a way of making sense of organizational dynamics. Strongly influenced by Anderson's (2006) analytic autoethnography, Sobre-Denton (2012) sought to transform her personal experience beyond self-observation by using analytical reflexivity of various occasions through three interpretive lenses of workplace bullying, cultural enactments of gender discrimination, and white privilege theory, employing the subsequent theoretical analysis of collected observations. Similarly, Van Amsterdam (2015) has used autoethnographic storytelling to share her experiences around breastfeeding in the academic environment by illustrating how the emotional and bodily reactions of colleagues convey the ways in which they perceive a newly maternal employee. Her detailed storytelling uncovered that lactating female academics encounter (and oppose) their own marginalization in a scholarly workplace environment.

Sambrook et al. (2014) examined how employee engagement can be studied from a more interpretive angle, arguing that autoethnography is an appropriate method to capture both the social and individual nature of self in employee engagement. Nevertheless, the authors of that study only identified the limitations of understanding the employee's self-concerning traditional disciplines, subsequently proposing a conceptual framework that considers the utility of autoethnography in studying the individual and dynamic nature of workplace engagement. A similar account of a particular struggle within the academic workplace has been undertaken by Pheko (2018), who investigated academic mobbing experiences related to the practice of power structures within academic institutions by sharing a personal exploration about the emotional and physical pain of autoethnographic research. The study offered an account of workplace bullying as a dysfunctional organizational behavior, detailing the cognitive response of an individual who had been subjected to bullying behavior by senior staff members bent on purposely obstructing the researcher's efforts to earn a promotion to the post of senior lecturer.

Popova (2018) conducted an autoethnographic reflection on her sexuality and decade-long employment in an organization that included supportive mechanisms toward LGBT diversity and inclusion. Her research has focused on understanding why and how sexuality-related issues have proven to be complex to tackle for both the author and human resource professionals within the workplace. Yoo (2020) used autoethnography to investigate motherhood in the context of academic environments by exploring her individual experience of work-life balance struggles. Her research aimed to frame issues faced by mothers grappling with the instability associated with working on a contract basis, the impact of accelerated academic time on mothering, and gauging the value of assembling a supportive community of other women to learn about similar mothering stories in academia and the struggles associated with them.

These past attempts to start a scholarly debate on the observed workplace-related subject, by and large, evolved around realism. They moved from story to interpretation, with researchers recording comprehensive and lively narratives and attempting to separate experience and the subsequent analysis. The following section will examine how realist autoethnographies can show and interpret a selected social phenomenon within a workplace chosen by identifying the research subject and the object of significance and analyzing the narrative connected with an existing theoretical framework.

2.4 Autoethnography as a methodological tool in the context of workplace research

As learned from described examples in the previous section, workplace autoethnography can be loosely classified under realist autoethnographies, in the sense that it uses the researcher's perspective or the perspective of other participants to establish a sense of authenticity by drawing upon personal accounts to describe and understand cultural experience through a complex narrative. Adams et al. (2015) noted that realist autoethnographies could take many forms. First, research reports and reflexive interviews enable researchers to use their experiences to complement and contextualize fieldwork. Second, ethno-dramas allow scholars to stage a live performance of participants' experiences and subsequently interpret them. Third, layered accounts permit autoethnographers to reflect upon the relationship between knowledge and interpretation by combining fragments of experience, introspection, research, and theory. Lastly, analytic autoethnography enables the researcher to acknowledge membership in a selected research community, reflect upon research experience collected through working in the field, and connect it with theoretical contributions to share a new perspective on the examined subject.

Since workplace research explores the connections between workplace design and the social aspects of human behavior, such as a worker's well-being, creativity, trust, and work organization (all of which tend to be connected to various forms of organizational structures), the workplace autoethnography requires a systematic inquiry approach to data analysis, commonly taking the form of an analytic autoethnography. As portrayed by Anderson (2006) and further adapted by Doloriert and Sambrook (2012), the latter requires several steps to fulfill the domain of analytic realism. It requires a complete member-researcher status in a selected group setting where the researcher needs to follow analytic reflexivity. Moreover, a prerequisite for this approach is the narrative visibility of the researcher's self, along with a visible narrative presence. The latter arises from the necessity of having a dialogue with informants beyond oneself and presupposes a commitment to an analytic research agenda focused on improving existing theoretical understanding of a selected social phenomenon – or, as summarized by Burnier (2006, p. 415), "[w]riting analytic autoethnography, the personal story plays an important, but a clearly subordinate role to the larger empirical-theoretical story central to analytic autoethnography."

Analytic autoethnography offers a means to revise and revisit a particular subject through the narrative of personal reality (McIlveen, 2008). The latter is especially useful in studies that examine scholar orientation and the social world as the approach. The subsequent analysis can demonstrate that an emerged assumption can be linked to the existing theoretical framework and conceptualization (DeBerry-Spence, 2010). The interpretative style of analytic autoethnography, where researchers' discourse is grounded in specifics from recorded reflections, tends to be suitable for research where the scholar is examining a particular physical setting. In other words, while evocative autoethnographies aim to observe personal emotions and feelings, the analytic autoethnographies have a goal to develop and refine the generalized theoretical understanding of a selected social process and enriching existing theoretical frameworks (Farrell et al., 2015), making it an appealing data-inquiry approach for workplace scholars.

2.4.1 *Defining the research process*

The following section proposes a structured approach when using analytic autoethnography as a method for data inquiry in workplace research, unveiling the methodological approach behind the conducted study. The core model behind an analytic workplace autoethnography has three phases. The fourth step suggests a circular process of repetition with a possibly similar research subject but a different object of signification.

Figure 2.1 portrays the analytic process of workplace autoethnography. The first step requires a scholar to identify the research subject and subsequently select the object of significance. As we have observed in the above narrative literature review where we have visited the existing attempts to understand a subject chosen in the context of a workplace, the identified research subject can focus on a wide range of topics, such as motherhood, sexuality, or employee engagement. Selecting the object of significance is essential to narrow down further the observed subject – the perception of breastfeeding by other actors in the workplace, a particular sexual orientation in the context of an office, or the conflicting nature of co-workers – in a selected scenario, to name a few examples that have been used in the past research.

The next step requires that the researcher narrates personal experience through the reflexive journey in a chosen workplace and context. Denzin (1989) has described personal experience narrative as a representational practice wherein social researchers take on the dual identities of their personal and scholarly selves, intending to tell autobiographic stories about a selected aspect of their lived experience in a particular context. The main challenge of a highly self-referential narrative practice is that it must be generated through a reflexive journey of researchers, capturing representations that are meaningful both to the author as well as to the text's audience, and informing the readers' understanding of selected aspects of the social world which necessarily exceed the autoethnographer's individualized experience (Butz & Besio, 2009). However, when carried out in a systematic, self-introspective way, the autoethnographic account enables the researcher to

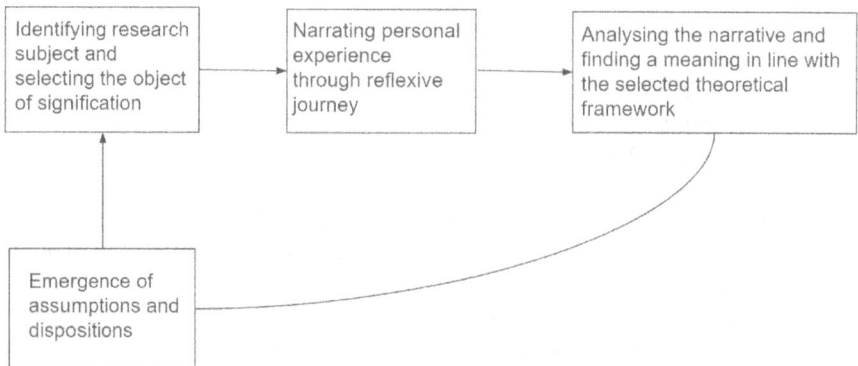

Figure 2.1 Analytic process of workplace autoethnography

analyze the connection to, actions within, and interpretations of, a selected setting (Maréchal, 2010; Thompson, 2015).

The third step consists of analyzing the recorded narrative and finding meaning in line with the selected theoretical framework. The next step would commonly look at the produced data using thematic analysis and frames the findings with an example, Positioning Theory, namely, to understand a scholar's position and his relationship to the transformation of the observed workplace. While thematic analysis enables us to identify common themes and representative patterns across the conducted (self) observations, Positioning Theory allows users to set an analytic framework, thus enabling the analysis of various aspects of identity through different subject positions. As Maydell (2010) observed, autoethnographers can engage Positioning Theory to produce a holistic representation of self as hinged from inside, as well as the formulation of identity as perceived by other actors. The following approach allows us to find a meaning and craft a story from the data that has been co-constructed by interacting with other individuals in a selected (work)space and a designated timeframe. Undertaking this reflexive journey commonly results in generating a distinct choice of meaning on behalf of the autoethnographer. The combination of analytic approaches facilitates a proper interpretation of the collected stories, and subsequently links them with relevant theoretical frameworks.

While the first step requires identifying research subjects and selecting the object of signification, autoethnographic observations commonly reveal new topics and uncover previously unseen objects of essences. With that in mind, analytic autoethnography is based on an evolving process which can be continued or interrupted at a certain point when researchers can assess that their reflexive journeys have captured a sufficient level of representations that can provide meaningful accounts.

2.4.2 The use of workplace autoethnography – A showcase

The following section shares a research journey of using flexible office space to replicate the daily workplace-related social processes that would usually commence within a traditional office in the pre-pandemic period, and their gradual transformation during the second wave of the Covid-19 pandemic. In that regard, the data has been collected in the total amount of about 200 hours throughout 32 periodic work visits of the selected coworking space. The successive report and analysis of the conducted autoethnographic study deliberately switch between the first- and third-person narrative to coherently (i) share a handful of autoethnographic stories that have been recorded in the line of selected research subject and changing objects of significance throughout a reflexive journey; and (ii) analyze these stories in relevant theoretical frameworks that enable to construct interpretations to events.

"While it is vital that the visibility of the researcher is maintained in the" narrative to capture the perspective transformations as they unfold throughout the reflexive process within analytic reflexivity" (Struthers, 2014, p. 189), self-observation around personal memories and embedded ethnographic accounts should not be generalized with broader implications for the shape of research around workplace practices. Rather, the autoethnographic narratives are contemplated in line with the contemporaneous body of social scientific knowledge relevant to contemporary workplace research. The further example is used as a showcase of how workplace autoethnography can be used. Due to the reporting of the stories and the subsequent analysis, the text switches from the third-person to the first-person perspective.

An exampled reflexive journey revolves around a personal transition from a home office resulting from the Covid-19 pandemic and a gradual decision to start using a local coworking space that had the potential to replicate both the office and the work-life balance that I had enjoyed back at the university. As a workplace scholar, I have been observing myself in the context of selected work environments for a handful of years and systematically started to record my experiences with an unforeseen transition to the home office:

> The sudden lockdown that happened as a response to the spread of the novel coronavirus has caught me unprepared like many other knowledge workers around the world. The first confirmed cases have caused the busy streets of the Czech capital city Prague to fall silent by mid-March 2020. People are shutting themselves behind the doors of their homes, and offices have gradually become inaccessible for most of us. The sudden transition has caught me off guard and, without my consent, transformed my home into a remote office – or, put another way, the only office from which I could work. My wife and I divided our one-bedroom apartment into several areas. Our Ikea dining table became my little office world. When we ate, I pushed my portable computer and a monitor on the side, and when we were done eating, the dining table underwent its transformation to a work desk. Days went by, cases were rising, and the suited news anchor constantly told me to stay home. I complied, stayed shut indoors and observed how the line between work and leisure was blurring. Frequent breaks and unmonitored performance put my productivity at serious risk. As a university employee, I had strict deadlines to follow, and the lack of motivation that I started to notice in the middle of April hasn't helped much. On top of that, my back started to hurt from prolonged sitting sessions on the kitchen chair. While the table did well when being used for dining, its transformation to a work desk resulted in ergonomic issues. By the beginning of May, I realized that my body and mind were in a gradual decline and that I needed a change. Upon returning to Prague, I found the university where I am employed still partially restricted in access, with my office turning out to be more-or-less inaccessible. A return to the home office was not an option, so I decided to opt for a local coworking space that is nested in the middle of my neighborhood. Being compact and having a maximum capacity of about three dozen individual users, I self-assessed that the use of that flexible office space would be relatively safe, signed up for a membership, and started to use it as my daily workplace.

The retrospective autoethnographic account summarizes the experience with the unexpected shift to a home office, and his perception of this shift in working arrangement. By the end of March 2020, about one-third of the world found itself in a lockdown, with corporations swiftly directing their office staff to work from home (Dubey & Tripathi, 2020). Most knowledge workers were forced to navigate the challenges of changing work environments, pondering how to optimally transform their homes into temporary office spaces (Waizenegger et al., 2020). The autoethnographic story records one of these challenging situations, where parts of an urban apartment were transformed into a workplace for a proportion of the time, and the struggle with ergonomic issues that have been extensively studied in the past (e.g., Juul-Kristensen & Jensen, 2005; Kogi, 2006; Törnström et al., 2008). Moreover, as self-observed and reported by Zhang et al. (2020) and Schieman et al. (2021), a sudden and unplanned transition to a home office can cause work-life conflicts, blurring the line between work and family life. The latter has become one of the main reasons to seek alternatives that would support remote work in a home-like environment.

Coworking spaces are shared and highly flexible workplaces that commonly offer various open or semi-open types of office spaces to accommodate knowledge workers pursuing individual or remote work arrangements (Rus & Orel, 2015). The first coworking spaces emerged in the mid-2000s as a direct response to the growing trend of self-employed individuals working from home or at a local café, facing challenges such as social isolation (Gerdenitsch et al., 2016), productivity issues (Bueno et al., 2018), and an appropriate work-life balance (Ivaldi et al., 2018). While experiencing a drop in the number of users due to the need for social and physical distancing in the first months of the pandemic, many coworking spaces managed to adapt until the end of the first wave by following strict health measures, and soon witnessed an increase of users whose employers retained a work from home policy (Gruenwald, 2020). However, it is not entirely clear how coworking spaces navigated the evolving constraints they faced throughout the pandemic; nor is it readily apparent how individual users changed their perception of an enclosed workplace shared with other persons who are generally not part of the same organization. The following autoethnographic accounts, therefore, provide insights into using a coworking space during the Covid-19 pandemic:

> Yesterday, I finally signed up for a six-month membership package that enables me to use the coworking space on a 24/7 basis. The community manager showed me around, briefly introduced me to a couple of people who were also using the workplace and signed me up on their Slack and other community-based channels. The space is bright, worktables are meant to accommodate two individuals at once, but it seems that everyone occupies the table by themselves as space is never at its full capacity. Two meeting rooms can be booked upfront and used for video calls that recently became the main way to communicate with my colleagues. On the first day, I don't have any concerns over the pandemic as the rate of new cases in the Prague-area is gradually slowing down, and it seems that people are taking social distancing precautions.

The narrative describes the selected coworking space and lays the groundwork for further exploration of the chosen subject. Coworking spaces generally have various types of packages that are determined by the time spent in the space. The majority of coworking spaces employ a community manager whose task is to forge a culture of collaborative interactions between coworking space users and form an internal supportive network of peers (Yang et al., 2019). Their actions mainly consist of directly or indirectly supporting the coworking space users and actively creating links between them using mediation mechanisms (Ivaldi et al., 2018; Rus & Orel, 2015). It is up to the individual users to capitalize on the networking opportunities that come with membership, and to self-identify with the internal network of support. The latter is frequently achieved by establishing active communication channels that keep an individual user informed about happenings within the coworking space, windows of formal opportunities, and contact with one another (Walden, 2019).

Coworking spaces are typically designed as open-plan offices with an abundance of desk that serve as workstations for several individuals. Spatial comfortability is essential for individuals' well-being, with the proximity of the workstations accelerating user engagement and possible cooperation between them (Orel & Alonso Almeida, 2019). Diversification of the space layout with private rooms and various furniture elements enable individuals to be flexible with their preferences toward workplace usage, which may vary from day to day. While the correlation between a coworking space's spatial design and the level of social interaction is well discussed, we are only now beginning to understand how coworking space users may react to the increased or decreased need for social distancing in the workplace designed to support a high frequency of interactions.

> Collective action is a common sight: due to the selected coworking space being a relatively small work environment, people are seated close together. The workspace appears to function as a catalyst for the community and their actions. A user gets a phone call, his phone loudly vibrates, he picks it up, and boomingly says 'Bullocks!' most of the other users start to giggle. It seems that there is a community dynamic for full-time users. Even when immersed in their work processes, they tend to share informal and personal experiences and collectively respond to them. I found myself doing somewhat the same, as it seems that I became prone to group dynamics. In another instance of group dynamics, there is a user who commonly does not mute his phone. There were already a handful of occasions where the user's phone rang. Other individuals raised their heads to express their non-verbal, collective disapproval of his failure to mute the ringtone on his phone. It is something that does not happen at home, but ironically, these distractions give me comfort, and I feel like I can work better. That said, I gradually started to notice that these common actions and responses positively affect my productivity and willingness to complete work tasks in a given timeframe.

One of the main goals of using autoethnography as a data-inquiry method is to capture and write about everyday interactions with others. But more importantly, "autoethnographers also describe moments of everyday experience that cannot be captured through more traditional research methods" (Adams et al., 2017, p. 4). The upper autoethnographic account captures the group engagement of individuals who frequent the selected coworking space, and, with it, the theme of collective action. Due to various social and spatial elements, they tend to form a coherent group of connected individuals prone to follow group patterns and expressions by engaging in shared reactions to a particular person's specific behavior or action. The richness of the autoethnographic narrative is particularly interesting due to the recorded impression of the autoethnographer. As Sambrook et al. (2014) argue when debating employee engagement as an individual concept, oneself is a highly personal dimension that fails to be captured in positivistic surveys. The latter can be even more evident in the context of a coworking environment as individuals are commonly not connected in the same organizational network but tend to replicate the level of interpersonal engagement that would be common for a traditional organization.

Blagoev et al. (2019) argued that coworking spaces show the capacity to frame and organize work, and, in doing so, provide the basis for collective action and patterning of work activities. These activities lead to what Garret et al. (2017) address as co-constructing a sense of community at work. Notably, three collective actions contribute to a sense of community: encountering, engaging, and endorsing. Coworking spaces tend to accelerate the interaction between individuals from various backgrounds by utilizing spatial elements of proximity (e.g., shared tables, common areas, etc.) or mediation techniques (e.g., community-orientated events) to construct a closely-knit network (Brown, 2017; Merkel, 2015; Rus & Orel, 2015). These efforts to promote interaction (and collective action) between coworking space users can create what Bouncken et al. (2021) call the workplace's "body language." Direct involvement in community-driven coworking spaces can directly affect an individual's self-motivation (Brown, 2017; Orel, 2019) and productivity (Clifton et al., 2022; Yang et al., 2019).

As observed on one occasion in late July, there are certain individuals who have a daily ritual involving a 'pull-up challenge.' They meet under a wooden frame with a metal bar placed in between the doors that lead to the meeting room and compete in the number of pull-ups, while discussing daily formal opportunities and bottlenecks (e.g., daily challenges with coding, dealing with demanding clients, etc.). This group of individuals appear as an informal elite, crème-de-la-crème of the coworking space, and it seems that you need to be invited to join, based on a referral. The group members generally interact with other users around other events, but the 'pull-up club' is their private little thing. One day, I randomly meet one of the guys from the pull-up club over at the coffee machine, where every member can get unlimited caffeine. We start to chat about this and find that we both share a passion for amateur boxing. A few days after, we bump into each other again at the same spot. I share my insight about the uppercut punch technique, and he is noticeably amused by my knowledge. In the early afternoon that same day, he approaches me and invites me to do some pull-ups with him and the others. I feel astonished that I 'got in,' and join without hesitation, do a couple of pull-ups, and discuss the market potential of a little-known cryptocurrency. My sense of belonging reawakens from hibernation, and I once again feel like I have a group of co-workers. At the same time, I wonder how long it will last.

Continued autoethnography revealed that inner workplace rituals link individuals who are otherwise not connected in an organizational network but work alongside each other daily or weekly. As Blagoev et al. (2019) observed, rituals provide the basis for coworking space users to connect and intimately share their work difficulties and personal issues. However, these rituals may also be used as a platform to share oneself with the supportive and sympathetic audience of peers where one can obtain an immediate response and aid. The invitation to attend one of those ritual sessions (i.e., the handlebar pull-ups) enabled me to see that the physical exercise stood for a bonding ritual where a selected group of individuals established an internal support network. The invitation to do my first set of pull-ups in a group setting initiated me into their inner circle. It served as a symbolic initiation, with which I have been informally accepted into a group.

Be that as it may, the continued pull-up sessions can be further seen as a ritual action that is "a form of social action in which a group's values and identity are publicly demonstrated or enacted in a stylized manner" (Islam & Zyphur, 2009, p. 116). Therefore, the ritual actions set a tone for the progress of a daily ritual by channeling the cognitive content, affective response, and behavioral activity toward the cultural expectations of the group's members. Furthermore, organizational rituals are commonly characterized as standardized and repetitive behaviors undertaken in conditions demanding direct performance expectations (Smith & Stewart, 2011). While rituals at work are habitually perceived as institutional learning and grouping mechanisms that build a healthy organizational culture (Erhardt et al., 2016), non-mediated gatherings such as the described "pull-up club" can be viewed as increased collaborative capability of a coworking space that evolves from the internal network's capacity for a collective action supporting innovation and performance in uncertain environments (Castilho & Quandt, 2017). The self-witnessed sense of belonging that I got after being initiated into a group provided me with the feeling of symbolical security, comradeship, and supportiveness that appears to be a needed foundation in times of disruption.

2.5 Implications and concluding thoughts

The presented narratives and subsequent analyses of first-hand accounts initially explore the author's relationship with a sudden change to the home-office environment, and subsequently investigate perceived work-productivity and well-being through a set of embodied experiences of how a local coworking space gradually transforms due to the progress of the Covid-19 pandemic. As is evident here, autoethnography can be used to shed light on different, often unseen aspects of workplaces. While more traditional ethnographic approaches could uncover some of the observed workplaces' elements, they would somewhat lack the ability to record researchers themselves and their own observations. These can greatly contribute to creating, and possibly contrasting, meaning with other individuals (or groups of individuals) who are actively or passively involved in a selected workplace.

The understanding of a particular workplace through stories enables the researcher to be conscious of their surroundings. In the presented case, the autoethnographer first uncovered the backdrop of network activities in the selected coworking space and discovered the patterns and rituals by actively participating in them and then reflecting on his experience. Second, the workplace autoethnography can help crystallize the inner organizational culture that tells researchers who they are by self-assessing the identification with space. Through the examined narratives, we could discern that the researcher was following the rising trend of territoriality in a non-territorial office space, mainly due to his internal fear of the Covid-19 disease, but also through his identification with the newly established norms to create a personal space within an office environment that is generally oriented toward close contact between individual users. Third, the researcher used the body's identity markers to confirm these trends and reflect upon the additional measures introduced following the rising number of newly discovered Covid-19 cases in the country. Lastly, the presented narratives enable others to understand themselves through researchers' stories.

Since the analytic and the presented form of workplace autoethnography offers a means to revise and revisit the subject through the narrative of personal reality, these stories and the accompanying analyses provide scholars with an orientation to conduct future studies and refine the generalized theoretical understandings of social processes. Knowledge work will see a gradual transformation within a post-pandemic society, with remote work becoming a new norm for many white-collar workers. While the coworking spaces and other forms of shared office environments would have undergone a swift popularization in the pre-pandemic world, the new and yet unimagined world of flexible work leaves many questions unanswered. However, the embodied experiences and the reflexive journey of a coworking space usage that has been recorded during (and between) two waves of the pandemic, and which has been successively analyzed, can give a framework of understanding as to how non-territorial offices can be impacted by social distancing and concerns of a contagious disease.

Despite the identified perspective of a workplace autoethnography, the latter would need to be tested and further developed in other settings and within a disparate context. This chapter has presented one of the possible analytical approaches to investigating allocated issues in a selected workplace setting. Nevertheless, evocative autoethnography could be another methodological framework that deserves more attention in the context of workplace research. This approach places more emphasis upon personal emotions and the researcher's feelings to uncover latent topics and matters in a designated workplace. And yet, interdisciplinary scholars might nevertheless seek the interpretative framework and theoretical contribution to the existing body of knowledge.

2.6   Further Reading

The reference list compiles some of rather fascinating reads on how to use autoethnography, especially Sambrook et al. (2014), Pheko (2018), Popova (2018), and Yoo (2020).

References

Adams, T. E., Ellis, C., & Jones, S. H. (2017). Autoethnography. In *The International Encyclopedia of Communication Research Methods* (pp. 1–11). Wiley.

Adams, T. E., Holman Jones, S., & Ellis, C. (2015). *Autoethnography*. Oxford University Press.

Anderson, L. (2006). Analytic autoethnography. *Journal of Contemporary Ethnography*, *35*(4), 373–395.

Blagoev, B., Costas, J., & Kärreman, D. (2019). 'We are all herd animals': Community and organizationality in coworking spaces. *Organization*, *26*(6), 894–916.

Bouncken, R. B., Aslam, M. M., & Qiu, Y. (2021). Coworking spaces: Understanding, using, and managing sociomateriality. *Business Horizons*, *64*(1), 119–130.

Bouziri, H., Smith, D. R., Descatha, A., Dab, W., & Jean, K. (2020). Working from home in the time of COVID-19: How to best preserve occupational health? *Occupational and Environmental Medicine*, *77*(7), 509–510.

Brown, J. (2017). Curating the "Third Place"? Coworking and the mediation of creativity. *Geoforum*, *82*, 112–126.

Bueno, S., Rodríguez-Baltanás, G., & Gallego, M. D. (2018). Coworking spaces: A new way of achieving productivity. *Journal of Facilities Management*, *16*(4), 452–466.

Buomprisco, G., Ricci, S., Perri, R., & De Sio, S. (2021). Health and telework: New challenges after COVID-19 pandemic. *European Journal of Environment and Public Health*, *5*(2), em0073. https://doi.org/10.21601/ejeph/9705

Burnier, D. (2006). Encounters with the self in social science research: A political scientist looks at autoethnography. *Journal of Contemporary Ethnography*, *35*(4), 410–418.

Butz, D., & Besio, K. (2009). Autoethnography. *Geography Compass*, *3*(5), 1660–1674.

Carroll, Á. (2020). An autoethnography of death and dying in Northern Ireland. Journal of Integrated Care, 28(4), 327–336.

Castilho, M. F., & Quandt, C. O. (2017). Collaborative capability in coworking spaces: Convenience sharing or community building?. *Technology Innovation Management Review*, *7*(12): 32–42.

Chong, S., Huang, Y., & Chang, C. H. D. (2020). Supporting interdependent telework employees: A moderated-mediation model linking daily COVID-19 task setbacks to next-day work withdrawal. *Journal of Applied Psychology*, *105*(12), 1408–1422.

Clifton, N., Füzi, A., & Loudon, G. (2022). Coworking in the digital economy: Context, motivations, and outcomes. *Futures*, *135*, 102439. https://doi.org/10.1016/j.futures.2019.102439

Cohen, L., Duberley, J., & Musson, G. (2009). Work–Life balance? An autoethnographic exploration of everyday home–Work dynamics. *Journal of Management Inquiry*, *18*(3), 229–241.

Cullen, J. G. (2011). Researching workplace spiritualization through auto/ethnography. *Journal of Management, Spirituality & Religion*, *8*(2), 143–164.

DeBerry-Spence, B. (2010). Making theory and practice in subsistence markets: An analytic autoethnography of MASAZI in Accra, Ghana. *Journal of Business Research*, *63*(6), 608–616.

Denzin, N. K. (1989). *Interpretative biography*. Sage.

Denzin, N. K. (1999). Interpretive ethnography for the next century. *Journal of Contemporary Ethnography*, *28*(5), 510–519.

Dolce, V., Vayre, E., Molino, M., & Ghislieri, C. (2020). Far away, so close? The role of destructive leadership in the job Demands–Resources and recovery model in emergency telework. *Social Sciences*, *9*(11), 196.

Doloriert, C., & Sambrook, S. (2012). Organizational autoethnography. *Journal of Organizational Ethnography*, *1*(1), 83–95.

Dubey, A. D., & Tripathi, S. (2020). Analysing the sentiments towards work-from-home experience during covid-19 pandemic. *Journal of Innovation Management*, *8*(1), 13–19.

Ellis, C., Adams, T. E., & Bochner, A. P. (2011). Autoethnography: An overview. *Historical Social research/Historische Sozialforschung*, *36*(4), 273–290.

Ellis, C., & Bochner, A. P. (2003). Autoethnography, personal narrative, reflexivity: Reseacher as subject. Collecting and interpreting qualitative materials, 2, 199–258.

Erhardt, N., Martin-Rios, C., & Heckscher, C. (2016). Am I doing the right thing? Unpacking workplace rituals as mechanisms for strong organizational culture. *International Journal of Hospitality Management*, *59*, 31–41.

Farrell, L., Bourgeois-Law, G., Regehr, G., & Ajjawi, R. (2015). Autoethnography: Introducing 'I'into medical education research. *Medical Education*, *49*(10), 974–982.

Gerdenitsch, C., Scheel, T. E., Andorfer, J., & Korunka, C. (2016). Coworking spaces: A source of social support for independent professionals. *Frontiers in Psychology*, *7*, 581. https://10.3389/fpsyg.2016.00581

Gruenwald, H. (2020). Coworking spaces in Germany during the Covid-19 crisis utilized for homeoffice and homeschooling. *South Asian Journal of Social Studies and Economics*, *8*(4), 57–67.

Hayano, D. M. (1979). Auto-ethnography: Paradigms, problems, and prospects. *Human Organization*, *38*(1), 99–104.

Holman Jones, S. (2016). Living bodies of thought: The "critical" in critical autoethnography. *Qualitative Inquiry*, *22*(4), 228–237.

Holt, N. L. (2003). Representation, legitimation, and autoethnography: An autoethnographic writing story. *International Journal of Qualitative Methods*, *2*(1), 18–28.

Islam, G., & Zyphur, M. J. (2009). Rituals in organizations: A review and expansion of current theory. *Group & Organization Management*, *34*(1), 114–139.

Ivaldi, S., Pais, I., & Scaratti, G. (2018). Coworking (s) in the plural: Coworking spaces and new ways of managing. In S. Taylor & S. Luckman (Eds.), *The new normal of working lives* (pp. 219–241). Palgrave Macmillan.

Juul-Kristensen, B., & Jensen, C. (2005). Self-reported workplace related ergonomic conditions as prognostic factors for musculoskeletal symptoms: The "BIT" follow up study on office workers. *Occupational and Environmental Medicine*, *62*(3), 188–194.

Kempster, S., & Stewart, J. (2010). Becoming a leader: A co-produced autoethnographic exploration of situated learning of leadership practice. *Management Learning*, *41*(2), 205–219.

King, P. (2019). The woven self: An auto-ethnography of cultural disruption and connectedness. *International Perspectives in Psychology*, *8*(3), 107–123.

Kogi, K. (2006). Participatory methods effective for ergonomic workplace improvement. *Applied Ergonomics*, *37*(4), 547–554.

Maréchal, G. (2010). Autoethnography. *Encyclopedia of Case Study Research*, *2*, 43–45.

Maydell, E. (2010). Methodological and analytical dilemmas in autoethnographic research. *Journal of Research Practice*, *6*(1), M5.

McIlveen, P. (2008). Autoethnography as a method for reflexive research and practice in vocational psychology. *Australian Journal of Career Development, 17*(2), 13–20.

Merkel, J. (2015). Coworking in the city. *Ephemera, 15*(2), 121–139.

Miller, K. (2002). The experience of emotion in the workplace: Professing in the midst of tragedy. *Management Communication Quarterly, 15*(4), 571–600.

Nagata, T., Ito, D., Nagata, M., Fujimoto, A., Ito, R., Odagami, K., Kajiki, S., Uehara, M., Oyama, I., Dohi, S., Fujino, Y., & Mori, K. (2021). Anticipated health effects and proposed countermeasures following the immediate introduction of telework in response to the spread of COVID-19: The findings of a rapid health impact assessment in Japan. *Journal of Occupational Health, 63*(1), e12198. https://doi.org/10.1002/1348-9585.12198

Orel, M. (2019). Coworking environments and digital nomadism: Balancing work and leisure whilst on the move. *World Leisure Journal, 61*(3), 215–227.

Orel, M., & Alonso Almeida, M. D. M. (2019). The ambience of collaboration in coworking environments. *Journal of Corporate Real Estate, 21*(4), 273–289.

Oswald, A. G., Bussey, S., Thompson, M., & Ortega-Williams, A. (2020). Disrupting hegemony in social work doctoral education and research: Using autoethnography to uncover possibilities for radical transformation. *Qualitative Social Work, 21*(1). https://doi.org/10.1177/1473325020973342

Pheko, M. M. (2018). Autoethnography and cognitive adaptation: Two powerful buffers against the negative consequences of workplace bullying and academic mobbing. *International Journal of Qualitative Studies on Health and Well-Being, 13*(1), 1459134.

Popova, M. (2018). Inactionable/unspeakable: Bisexuality in the workplace. *Journal of Bisexuality, 18*(1), 54–66.

Reed-Danahay, D. (1997). *Auto/ethnography: Rewriting the self and the social.* Berg.

Riad, S. (2007). Under the desk: On becoming a mother in the workplace. *Culture and Organization, 13*(3), 205–222.

Rus, A., & Orel, M. (2015). Coworking: A community of work. *Teorija in Praksa, 52*(6), 1017–1038.

Sambrook, S. A., Jones, N., & Doloriert, C. (2014). Employee engagement and autoethnography: Being and studying self. *Journal of Workplace Learning, 26*(3/4), 172–187.

Schieman, S., Badawy, P. J., Milkie, M. A., & Bierman, A. (2021). Work-life conflict during the COVID-19 pandemic. *Socius, 7*. https://doi.org/10.1177/2378023120982856

Schur, L. A., Ameri, M., & Kruse, D. (2020). Telework after COVID: A "silver lining" for workers with disabilities? *Journal of Occupational Rehabilitation, 30*(4), 521–536.

Smith, A. C., & Stewart, B. (2011). Organizational rituals: Features, functions and mechanisms. *International Journal of Management Reviews, 13*(2), 113–133.

Sobre-Denton, M. S. (2012). Stories from the cage: Autoethnographic sensemaking of workplace bullying, gender discrimination, and white privilege. *Journal of Contemporary Ethnography, 41*(2), 220–250.

Spry, T. (2001). Performing autoethnography: An embodied methodological praxis. *Qualitative Inquiry, 7*(6), 706–732.

Struthers, J. (2014). Analytic autoethnography: One story of the method. In *Theory and method in higher education research II.* Emerald Group Publishing Limited.

Thompson, J. D. (2015). Towards cultural responsiveness in music instruction with black detained youth: An analytic autoethnography. *Music Education Research, 17*(4), 421–436.

Tillmann, L. M. (2009). Speaking into silences: Autoethnography, communication, and applied research. *Journal of Applied Communication Research, 37*(1), 94–97.

Törnström, L., Amprazis, J., Christmansson, M., & Eklund, J. (2008). A corporate workplace model for ergonomic assessments and improvements. *Applied Ergonomics*, *39*(2), 219–228.

Van Amsterdam, N. (2015). Othering the 'leaky body'. An autoethnographic story about expressing breast milk in the workplace. *Culture and Organization*, *21*(3), 269–287.

Waizenegger, L., McKenna, B., Cai, W., & Bendz, T. (2020). An affordance perspective of team collaboration and enforced working from home during COVID-19. *European Journal of Information Systems*, *29*(4), 429–442.

Walden, J. (2019). Communicating role expectations in a coworking office. *Journal of Communication Management*, *23*(4), 316–330.

Wall, S. (2006). An autoethnography on learning about autoethnography. *International Journal of Qualitative Methods*, *5*(2), 146–160.

Yang, E., Bisson, C., & Sanborn, B. E. (2019). Coworking space as a third-fourth place: Changing models of a hybrid space in corporate real estate. *Journal of Corporate Real Estate*, *21(4)*, 324–345.

Yoo, J. (2020). An autoethnography of mothering in the academy. *The Qualitative Report*, *25*(8), 3173–3184.

Zawadzki, M., & Jensen, T. (2020). Bullying and the neoliberal university: A co-authored autoethnography. *Management Learning*, *51*(4), 398–413.

Zhang, S., Moeckel, R., Moreno, A. T., Shuai, B., & Gao, J. (2020). A work-life conflict perspective on telework. *Transportation Research Part A: Policy and Practice*, *141*, 51–68.

3 Affective ethnography

Interpreting body language in the workplace fieldwork

Lotte Holck and Florence Villeseche

Copenhagen Business School, Denmark

3.1 Introduction

This chapter explores the role of bodies and their related affect hitherto often ignored in the ethnographic fieldwork at the workplace (Cunliffe & Karunanayake, 2013; Gherardi, 2019; Gilmore & Kenny, 2015; Holck, 2018a; Holck & Muhr, 2020; Manning, 2016, 2018). Thus, the body has become taken for granted as 'absent presence' within workplace ethnography (Ashcraft, 2018; Pullen & Rhodes, 2015). As the diversity in and around workplaces is increasing, the adoption of workplace ethnographic research is becoming a more appropriate – but also more challenging – methodological approach. In diverse workplaces, the internal complexity related to employees' race, gender, culture, and language adds to the more conventional workplace differentiation of tasks, functions, positions, team and section structures, communication lines, organizational (sub-)culture, hierarchy, or power distribution (Acker, 2011 Muhr et al., 2022). Organizational members more tacitly and implicitly enact demographic bodily features and relations. Accordingly, issues of the body and bodily relations make the research of workplace ethnographers much more complex, multi-layered, and intersubjective, distorting and contradicting the ideal of neutrality and objectivity embedded in (traditional) ethnographic research (Cunliffe, 2010; Hibbert et al., 2014; O'Reilly, 2012; Van Maanen, 1991, 2011).

My study demonstrates how bodily features such as age, skin color, or national background impact practices of collaboration and workplace relations as well as researcher-participant relations. This is inspired by my experiences of doing janitor work as a middle-aged Danish woman working alongside young employees with predominantly migrant backgrounds, and how this impacted my relations with 'colleagues' and the data I was generating. As my younger colleagues with migrant backgrounds had to train me as a new employee, the 'ordinary' hierarchical relations were twisted and disturbed as native middle-aged Danes usually take up managerial positions while migrants habitually occupy lower subordinate frontline positions. The bodily features and interaction of and between researcher-research subjects are thus important factors to bring into consideration; also to make meaningful the kind of data we generate together and the eventual findings. My claim is that this is not only relevant for my particular

DOI: 10.1201/9781003289845-3

study, but for workplace ethnographic research and research in diverse organizational settings in general.

This research was taken up on the initiative of the HR director in my case company, the Danish service company Clean.[1] Clean is generally perceived as a diversity champion in Denmark due to its high degree of employee diversity and company values of diversity and inclusion. The HR director wanted to learn more about the way team diversity impacts – and, preferably, enriches – collaborative and leadership practices in diverse teams[2] in Clean. As a researcher of workplace diversity and inclusion, I was asked to do research on the subject and consequently help the HR manager and the company to identify local best practices to help increase the number of diverse teams, which was the goal of Clean. Team diversity is officially communicated to boost innovative thinking in teams, avoid damaging ingroup-outgroup relations among different (ethno-racial) subgroups, and strengthen leadership skills in Clean. I was given almost free hands in terms of fieldwork methods. As an urban geographer by training, I am inspired by classical workplace ethnographies from the legendary Chicago School (see Silverman, 2001; Yanow, 2012; Zickar & Carter, 2010) and wanted to 'go native' to explore the participants' daily interactions and relations by entering several diverse janitor teams to work alongside and collaborate with the team members with a 'progressive change agenda' in mind.

However, to my surprise, I realized that my fieldwork methods were like "putting a spoke in the wheel": I could not enter the teams unnoticed. Slowly I realized how my engaged ethnographic methods distorted the 'taken-for-granted' and revealed the more tacit, implicit hierarchical routines and relations. An 'inequality regime', as coined by Acker (2011), started to emerge, that is an organizational power landscape consisting of both formalized, explicit structures of equality (e.g., a formalized diversity policy in Clean) and more informal, subtle substructures of inequality (see also Holck, 2018b). The latter are often tacitly practiced and through the mundane daily organizational interactions among employees, in which, for example, gendered, age-related and/or 'ethnified' assumptions about minority/majority are enacted and inequality (re)inforced (Acker, 2011, see also Ahmed, 2014; Muhr & Sullivan, 2013; Romani et al., 2019). These substructures of inequality can lead to systematic disparities between employees regarding power, control over goals, resources, and outcomes; decisions on how to organize work; opportunities for promotion and interesting jobs; security in employment; and so on (Acker, 2011). As I will return to below, this was not the kind of findings that the HR manager – nor I – had anticipated or hoped for.

3.2 Background

Workplace ethnography has traditionally relied on an interpretative framework that necessitates the researcher to 'go native' to take the participants' perspective (O'Reilly, 2012; Van Maanen, 1991). From the early days of the Chicago school, organizational researchers capitalized on the rich field of workplace ethnography, using ethnographies to inform their research and provide a sense of the realities of the workplace by applying a vast array of methodological techniques, including sociometry, case studies, interviews, and intensive observations (Smit & Onwuegbuzie, 2018; Zickar & Carter, 2010). The ambition of the early workplace ethnographers was to study workers by working alongside them to get a sense of the 'reality' of the workplace, instead of treating workers as abstract entities like many economists did. Hence, these methods draw from the original in-depth approach of the cultural anthropological tradition, where observing the research subjects in their habitat is central (Clifford, 1983). The ambition of the workplace ethnographers was reformative, as they wanted to understand the conditions of the working lives of the working class by entering the workplace and use their findings to mitigate the poor conditions of the urban industrial workers of the early twentieth century (Zickar & Carter, 2010).

Traditionally (and to some degree still), the term 'going native' has been used to describe the way the researcher ideally should become 'one' with the daily life of the research subjects. A central characteristic of ethnographic writing is the ability to convey the sense of what can be known about organizing processes by 'being there'; "being immersed in the situations, events, interactions, and so forth that provide the grist for the ethnographer's knowledge claims mill" (Miettinen et al., 2009, p. 1316). Ethnography is traditionally seen as the result of the ethnographer's efforts to soberly describe what he or she experiences through immersive, lengthy participant observations in the field (Smit & Onwuegbuzie, 2018). The data collected in workplace ethnography often results in a combination of autobiographical accounts of and interviews with the participating employees, work diaries, and participant observation in extensive fieldwork notes on the researcher's behalf. This qualitative data can be complemented with organizational data like statistical records, salary structures, organigrams, company documents on values, policies and strategies, Employee Satisfaction reports, etc. (Silverman, 2001). The data collected during ethnographic work is then analyzed in a holistic manner. This involves 'thick descriptions' focused on detailed empirical data together with an interpretive effort that goes beyond or beneath specific manifestations by interpreting layers of meaning (Van Maanen, 2011). Ethnography requires both immersion and distance to translate experiences and make them meaningful to the reader (Cunliffe & Karunanayake, 2013).

Workplace ethnographic research has long been challenged. First of all, the replicability of quantitative studies has been promoted as they are much easier to conduct and require less personal time and resource investment on behalf of the researcher and the case organization (Zickar & Carter, 2010). Second, a concern for the researcher's personal bias has thwarted the ideal of workplace ethnographies as a "highly descriptive writing about particular groups of people" (Silverman, 2001, p. 305). Third, postcolonial researchers have criticized ethnography as born out of the colonial desire to observe and understand the colonized 'Other', which gives the term 'going native' an increasingly troublesome connotation as a problematic practice and an offensive label (Holck & Muhr, 2020). Postcolonial researchers underline how when conducting ethnographies, you need to ensure the "delicate balancing act of empathy and distance" (O'Reilly, 2012, p. 89). Consequently, this approach demands self-reflexivity on behalf of the ethnographer to make "knowledge claims but to make them with a self-awareness of her personal politics, privileges, and resources" (Dar, 2018, p. 571). Self-reflexivity is the capacity to recognize that the particular tradition of the researcher mediates all accounts of the workplace, which methodologically and epistemologically challenges the objectivity, neutrality, and scientism pervading traditional ethnography (Flores, 2015; Gilmore & Kenny, 2015). As such, context and subjectivism are vital parts and parcels of the research process that must be scrutinized through self-reflexivity (Hibbert et al., 2014).

Self-awareness in terms of the researcher's politics, privileges, and resources becomes particularly important in ethnographic research in diverse settings, and even more so if carried out by white, Western, or majority researchers (Manning, 2016, 2018), as it is impossible to erase a white researcher's bodily features that are also markers of politics, privileges, and resources (see also Flores, 2015; McCorkel & Myers, 2003). According to feminist and postcolonial scholars, difference markers like race or gender are constructed performatively: bodies are never neutral but gendered and 'ethnified' through historic and culturally produced categories of difference (Ahmed, 2000, 2004, 2007, 2017; Andreassen & Myong, 2017). The 'colonial heritage' of ethnography increases the awareness of the importance of self-reflexivity through the researcher's self-examination and awareness of positionality to account for the data-generating process (Cunliffe & Karunanayake, 2013; Holck & Muhr, 2020). Critical scholars have a profound skepticism regarding the possibility of an objective and disinterested foundation for knowledge. Instead, the impossibility of methodological engineering that separates "the knower from what is known" is highlighted; workplace ethnography must embrace the perception that all knowledge is socially constructed (Cunliffe, 2010; Hibbert et al., 2014).

3.3 Argument

This chapter argues that workplace ethnographic research has been more or less oblivious to the role of the researcher's body and bodily sensations in interactions with the research participants. This is despite the current debate on self-reflexivity among workplace ethnographers, who acknowledge their immersion in a specific, historically contingent research setting (Cunliffe & Karunanayake, 2013; Hibbert et al., 2014; Miettinen et al., 2009). But the notion of self-reflexivity in itself does generally not pay attention to the researcher's body and bodily sensations when generating and interpreting data from the field.

The embodied phenomenology of Merleau-Ponty has been highly influential in articulating the relationship between perception, body, and how we make sense of the world (Shilling, 2012). Merleau-Ponty highlights how perception and cognition happen in and through our bodies, and precedes the processing of experiences by consciousness. This means that perception is an embodied experience that occurs in the world. We know the world not through our intellect but through our bodily experience of sensing emotions and meanings in people's facial expressions, gestures, and the rhythm of action (Mikkelsen, 2022). The perceiving body is not a passive receiver of 'messages', but an active experiencing and interpreting body that is practically engaged with its social setting. Accordingly, fieldwork data generation (also) happens through the body and bodily sensations of social relations and events that we might not understand intellectually. However, we still appreciate intuitively through pre-reflective sense-experience of sight, sound, and touch (Mikkelsen, 2022) as we are present in-the-world/the workplace.

A way to empirically grasp the relation between body and perception, and hence the role of bodies and bodily relations in research, is to pay attention to affect. According to Ahmed (2004, 2014), affect focuses on bodily reactions and relations; affect is the space between bodies and mediates between bodies as a kind of bodily energy, making affect contagious and dependent on relations between interacting bodies (Ahmed, 2007; Holck, 2018a). Affect does embrace not only a single body's expression but also the whole mood and atmosphere of the relational situation (Ahmed, 2014). The mood is a socially situated phenomenon, it is not something we possess but something that takes hold of us (you are 'in a mood', see Ahmed, 2014). A person's mood can be attuned to, out of tune, or even misattuned with the situation summing up to the collectively experienced feeling of the social situation (Ahmed, 2014, 2017). Misattunement is experienced as being 'out of sync' and 'troublesome', creating the figure of the outsider or the stranger spoiling 'the good atmosphere' (Holck, 2018a). Misattunement can be caused by the arrival of bodies that do not 'fit' the situation (Ahmed, 2014) – in my research, it was the arrival of a middle-aged majority group body that was to perform janitor work alongside younger employees with migrant backgrounds.

In the next section, I will establish how dealing with my bodily sensations and my bodily interaction with the research participants, when doing fieldwork in Clean, influenced the research situation. The examples below highlight how the researcher's affective bodily reactions and sensations together with the observed persons' bodily responses must be considered when accounting for the generation and interpretation of data.

3.4   Example of application/use

As mentioned above, the purpose of my research in Clean was to explore the impact of the difference in collaboration and leadership in diverse teams in Clean. Clean is a major employer in Denmark and is publicly renowned as a 'diversity champion'. The approximately 7000 employees come from 118 different countries. 50% of the employees have migrant backgrounds (predominantly from Eastern European and non-Western countries), and all work at the lowest organizational level among frontline managers and employees (86%).

Drawing on engaged workplace ethnographic fieldwork methods involving participant observation, I was trained as an apprentice to work alongside janitors in 20 diverse teams during six months in Clean. The official definition of a diverse team in Clean is a max. of 70% homogeneity in terms of gender, generation, and racial background among the team members, which also made up the criteria when selecting the relevant 20 teams for me to do fieldwork in. However, in practice some of the teams were not diverse in terms of gender and generational differentiation; half of the teams were made up exclusively of younger women. But most teams adhere to ethnic diversity and would consist of a mix of people from different countries. The teams would be different in size from ten up to 35 team members. The HR department conveyed contact and access to the relevant teams. However, the HR department was unaware of which teams I eventually ended up working with, and the names and geographic locations have been anonymized to protect the participants. See Table 3.1 for an overview of my data in Clean.

I would turn up at five in the morning, face wrinkled from too little sleep and the early bike ride in my acrylic janitor uniform two numbers too big, to be trained in the twenty teams, by different janitors each time. In eighteen out of the twenty teams I spent two to three days, but in two teams I was luckily able to work with

Table 3.1 Overview over data and fieldwork in Clean

Elements	Data	Dates
Observations and interviews	Ethnographic observations of teams while being trained as a new employee for two days in each team, in a total of 18 teams all over Denmark	April 2016 to June 2017
	Two in-depth case studies of 15 days observation at 2 larger sites	
	Two focal interviews each with six team leaders and contract managers both in West and East	
	Shadowing of 20 team leaders and supervisors	
	20 semi-structured interviews with leaders at different levels of the organization	
	Daily field notes and transcription of interviews and teaching sessions	
Interventions	*Content*	*Dates*
	Six presentations of results internally for members of the HRM department and the regional leadership team of the cleaning division and a final conference presenting the findings including external practitioners	April 2016– Oct 2018
	Teaching at four internal leadership development courses for team leaders and supervisors on "diversity management of diverse teams" and development of curriculum	
	Participation in seminars, social events, debates, and informal conversations in Clean	
	One internal report midway and four internal reports communicating main findings within key areas of concern ordered by Clean (HR director)	
	Reciprocal meetings with members of the HR department, HR director, and the director of the communication department	

them for an extensive period of two weeks, which helped to deepen my experience of the working of diversity in teams. I also shadowed the managers of those
twenty teams. I furthermore interviewed twenty leaders at different levels in the
administration of Clean (predominantly from communication, site and key account
managers, and the HR officers). The full data set is not used in this chapter, even
though it holistically shaped my fieldwork experience in Clean.

I present the data as two vignettes based on what I observed and experienced
while in the field, dwelling on my own bodily sensations and affective reactions.
Presenting the data through vignettes supports the focus on micro-interaction vital
to explore bodily sentiments and interaction (Holck & Muhr, 2020; Miettinen et al.,
2009), as the body and bodily relations are not possible to grasp in interviews only.
Even if they were, presenting the data through interview excerpts quickly loses the
focus on micro everyday activities and comments making up the feeling of 'being
there', sensing and perceiving in and through the body 'what is going on'. Through
vignettes, I am able to pass on the richness of the data, several people's behavior, (re)actions, and the dynamic between them – and me – in the work situation.
However, all text in the vignettes is based on what I experienced regarding social
relations and what I observed research participants say or do.

This method demands extensive fieldnotes; every day after 'work' I would
spend about three hours meticulously writing detailed field notes. They would be
my accounting for the whole day from arrival to departure; whom I met with and
collaborated with; participants' comments on their daily work, team collaboration,
and reactions to the task to train me as an apprentice. I would pay attention to their
bodily reactions and facial expressions when collaborating with me in a pair or
in social situations when the whole team was present during meetings or breaks.
Following the notion of affect, I would pay particular attention to social situations
and interactions where I experienced instances of bodily reactions of awkwardness,
embarrassment, or shame; feeling hot or cold, blushing, shaking, sinking feeling
in the stomach, etc. These instances I interpret as affective moments, where my
bodily reactions revealed feelings of being an intruder in or invader of the participants' working space, of being 'in the wrong'. In the fieldnotes, I focus on describing affective situations, where I as a researcher felt that I 'spoiled' the collective
mood, when I felt misattunement or the misfit of my body in the social setting.
This also includes attention to the bodily reactions of the research participants,
as they display bodily responses of, for instance, confusion, embarrassment, and/
or exclusion related to their bodily social interaction with me; turning their backs,
not responding to my questions, clipped voices, long silences, and 'telling' gazes
between colleagues.

The lens of the body and affect highlight two related 'bodily troubles' that I
encountered and which have implications for my way of engaging with the fieldwork: First, my urge to 'pass' and experience what it feels to be 'insider' while
constantly being reminded of my 'ill fit' and outsider standing. Second, the impact
of affective attunement, as I highly desired to be included yet my demonstrations
of solidarity with the participants was met with gestures of exclusion. These two
examples of 'body trouble' will be dealt with in the following two vignettes to

demonstrate how attention to the body and affect can be highlighted within workplace ethnographic research.

3.4.1 Story one: The misfit body

Phueng is my trainer for the day. She giggles when the site manager Lise tells her that she has to train me for the day. "Are you Polish?" giggling, her hand covering her mouth. "No I am Danish. Are you Danish?" More giggling. "No. But you are Polish, right?" I realize that Phueng cares about my whiteness, my age, and my nationality at six in the morning, and it is too cumbersome to explain what I am doing and my demographic background. After a while her colleague Abir arrives, she stares at me and whispers with Phueng. Abir asks me "Are you Polish? You look like a Polish lady". During lunch in their team, I am introduced as a new Polish employee. Cleaning in the tea kitchen in a big office building, a customer – a woman in a piece suit my age – stares at me: "Are you new?" "Yes I am training as a new employee". "You are doing soooo good", she exaggerates her expression, speaking slowly with a hand on my shoulder. Does she believe I am slow or mentally challenged? Does everybody think like this? Is that because of my whiteness and age, being trained by a young janitor with Asian background? These instances make me wonder about how others perceive me – customers and janitors – when I am wearing a uniform. The frequent questions of "why are you here? Where are you from?" make me uneasy and painfully aware of the misfit of my body when working in the janitor teams. I grow increasingly aware of how exactly the noticeability of demographic characteristics of my body makes it difficult for me to 'pass', as I am both highly visible and invisible: While janitors in one team settled with calling me 'the Polish lady', others explicitly 'ignore' my presence by speaking other languages than Danish or English. I slowly come to terms with how my bodily features – being white, female, Danish – creates a 'queering effect' of confusion and awkwardness (Ahmed, 2007).

Training as a janitor was hard on the body in terms of heavy, monotone manual work at night or early morning shifts. Reading through my field diaries, the dominating themes are not on the physically but the mentally straining work of constantly trying to fit in, passing as 'one of the janitor' teams, and actively interpreting situations with colleagues and customers. My fieldwork diaries are prone with reflections registering my outer and inner bodily sensations when working alongside the janitors. My inner voice going through the field diaries express an urge to expound and process gestures and moods related to my own misunderstandings, feelings of clumsiness, and 'affective vigilance': I constantly felt like 'being in the way' of effective work performance, felt watched and categorized in a way that swayed my self-perception and room for maneuver in the research situation. These fieldwork feelings and experiences contradicted my original intention of 'going native' (Van Maanen, 2011) – in fact, it was impossible for me to pass as a native; I was an outsider in a problematic and misplaced body.

In my fieldwork, reflections of my bodily sensations gave me insight into two dimensions of my fieldwork: On the one hand, I had prominent feelings of my body as a 'misfit', a 'wrong body' among janitors. These sensations highlight how the interplay of intersecting demographic bodily features are essential in terms of which bodies are perceived as 'right' and 'wrong' to do different work in an organization. This is in line with observations made by critical scholars on how bodily characteristics such as ethnicity, nationality, age, class, and gender are explicatory to practices of organizational stratification (Acker, 2011; Ahmed, 2000, 2004, 2007, 2014; Andreassen & Myong, 2017; Ashcraft, 2013; Dar, 2018; Flores, 2015; Holck & Muhr, 2020; Kennedy-Macfoy and Nielsen, 2012; Manning, 2016, 2018; McCorkel & Myers, 2003). These instances thus deal with how my exterior/physical body traits created assumptions about my identity that predetermined and/or shaped my possible room of maneuver in the fieldwork situations. My body and bodily features are thus formative to and embedded in my relations with fieldwork participants.

Second, I experienced 'inner' bodily sensations and affective impulses in research situations that influenced my interpretation of situations and observations made in Clean. These redirects focus from 'objectively' scrutinizing the collaborative situation to a much more subjective focus dwelling on the impact of the simultaneity of both my 'exterior' and 'inner' bodily sensations and what this 'did' to me. My urge to pass and experience what it feels like to be an 'insider' was being replaced with a focus on my 'ill fit' and outsider standing. It took me a while to gather that it was not because I was bad at doing research, or using the wrong data-generating methods. I discovered how doing engaged, participative research always involves the researcher's body and bodily relations with participants. Noticing these inner bodily reactions and the social mood of the research situation makes data more complex and messy. Accordingly, researchers often ignore them as irrelevant or as potentially spoiling the neat and logical presentation of data for publication.

3.4.2 Story two: The excluded body

Working alongside people day in and day out makes you want to fit in or at least be accepted. When being in the field, I – to my own wonder – started longing for signs of acceptance and inclusion among the janitors. However, I constantly felt like an intruder, an outsider. The longer I stayed in the organization, the more I felt like a stranger and outsider, never experiencing being 'fully included'. A few times, I experienced a small sense of 'insiderness': When I asked for the canteen the first day in a team and colleagues warned me "You wouldn't use that, it is too expensive" and shared their food with me. Or when colleagues spent their breaks telling me their 'stories' and about their work, or helpfully instructed me. But eventually these instances of 'insiderness' were disrupted to remind me of how I was indeed not considered an insider. For instance, at lunch break we were interrupted by the site manager, Jens, who asked me "are they treating you well?" followed by an awkward silence in the team of janitors. Waves of shame rushed through my body accompanied by thoughts like: "Do they think I am a spy for management? A traitor telling the manager about their plot?" They had just been talking about how to help cover their pregnant colleague Aisha's work without involving the site manager. The rest of the lunch was not as cheerful and the food gracefully shared was stuck in my throat. At other times, groups of colleagues would start to speak Turkish, Thai, or Romanian during breaks, which made me acutely aware of their shared identity, excluding me as I was not understanding a word. During working hours, I was treated differently to other 'apprentices'. Sometimes, colleagues made me clean all the men's toilets, as I as an outsider should do the most 'shitty' work while reserving the better tasks – cleaning offices – to the insiders. Others were afraid to make me work at all and with great embarrassment, I watched them do all the hard work while I was left with small easy tasks. All to my disappointment. I was constantly wondering how to gain their confidence: "Can I ever know what it is like to be 'in their shoes'? And what does that exactly mean?" My body constantly created an unrest and a stir among the research participants: my white Danish body was affiliated with leadership, and I was constantly visible with my white body among young, colored bodies. This made me stick out as a strange and excluded body when all I wanted was to fit in to get a sense of 'insiderness'.

As mentioned, my study was originally inspired by the idea of the ethnographer going native, to immerse herself through longitudinal research and become a quasi-insider. However, the situations where I was supposed to generate data, observe participants, and ask questions were often reversed as I felt interrogated and watched. This was, to me, a disruptive change in researcher-research relations that I had not anticipated. Naïvely, I had thought of many other fieldwork problems like gaining access and collecting relevant data through time-consuming participatory observations. During my stay in Clean, I had not foreseen this problematization of my body's presence and my very affective bodily reactions and sensations.

"Not to inhabit the norm (or not quite the norm) can be experienced as not dwelling so easily where you reside. You might be asked questions and made to feel questionable so that you come to feel that you do not belong... you might turn up and not be allowed in or find it too uncomfortable to stay!" (Ahmed, 2017, p. 115). My body was constantly made into a question mark and did not seem to belong. When your body is not perceived to belong, you will be asked questions to account for yourself: "where are you from?" or "why are you here?" is a way of being told that you do not belong (Ahmed, 2017). Hence, you must 'explain' yourself understandably and legitimately to the 'inquisitor'. In Clean, I had to account for how I ended up as a white person in an otherwise 'brown' person occupation, as whiteness is affiliated with leadership and white-collar work, involving sitting at a desk. Whiteness was registered as ill-fitting, foreign, and not to belong among the janitors. The questioning and stopping of my body not passing the norm, creating a strong sense of being out of place like the ill-fitting uniform dressing my body. Indeed, any form of fieldwork can lead to feelings of doubt, frustration, and confusion as the researcher struggles with choices about the nature of relationships in the field, multiple sources of data and what is or isn't relevant, ethical dilemmas, and what this all means for theorizing and writing (Cunliffe & Karunanayake, 2013; Holck, 2016, 2018a).

However, my bodily feelings and the urge to belong or somehow fit in while researching puzzled me. According to Merleau-Ponty and Smith (1962), we *are* bodies – we are not merely having a body – which inevitably draws attention to affective sentiments and experiences rushing through the body as important while in the field. We sense through our bodies, pick up the mood, and not fitting in is highly uncomfortable. You become painfully self-aware, monitoring the situation, other bodies' reactions and sensing your own body getting hot, sweaty, or stiff. The affective body reactions take up much energy in the research situation.

The same goes for the other bodies present. It is not an individual feeling: the body is always collective, it is more than just one body (Manning, 2016). The 'collective of bodies' highlights the contagious aspect of interacting bodies when doing research. It is never an individual body but a body in interaction with other bodies creating affective, collective, social feelings that manifest as 'mood' shared among the research participants (Gherardi, 2019). My bodily uneasiness and feelings of 'misattunement' were contagious as the participants more or less intuitively sensed my moods through pre-reflective sense-experience of sight (my facial expression of confusion or shame, my clumsy movements in awkward situations), sound (my voice shaking or falling silent), smell (of my nervous sweating), and touch (bumping into each other, the 'electric impulse' when hands touch when greeting or handing things to each other).

Dwelling on these affective bodily reactions when doing research helped me to discover layers of meaning of 'what is going on': my attempts to affectively attune my body to the social situation when working in the janitor teams; my desire to be included and demonstrate solidarity while experiencing exclusion and the questioning of my body; these sentiments are ways to explore the boundary drawing process of inclusion and exclusion that (re)produce team collaboration. I did not experience how diversity was enacted in the teams in Clean as a native or insider trying to incarnate the participants' perspective (van Maanen, 1991). Instead, my position as an outsider invoked instances of affective engagement with the participants, which served to demarcate the boundaries of inclusion and exclusion in the team.

3.5   Implications

3.5.1   *Method relevance to research*

As shown above, a focus on bodies and affective bodily relations can fruitfully be applied in workplace studies in general and in particular when inquiring into diversity and inclusion. As diversity is increasing in and around workplaces, doing more research on workplace inequality and inclusion is most needed. With a focus on my body and bodily relations with research participants, my study revealed issues of 'ethnification' of work and positionality (Holck, 2018a; Zanoni and Janssens, 2015), and the dynamics of boundary drawing in terms of inclusion and exclusion in teams. These findings did not immediately come up while in the field, where it rather felt like a nuisance to 'stick out' with my white middle-aged body, not to be included as an 'insider'. But as I started dwelling on my awkward fieldwork experiences – and got past feelings of humiliation and shame – I started interpreting layers of meaning embedded in these situations. I discovered how these experiences connected to subtle social and demographic dynamics in Clean that would be difficult to reveal through qualitative methods.

A focus on the researcher's body and bodily sensations brings about two important dimensions of ethnographic research at the workplace: First, sensitivity to all the aspects that influence our research when we generate data and analyze our findings, since there is no unprejudiced, neutral, or unbiased access to research. As a researcher, I was a co-constructor of the empirical data on which I based my findings. Furthermore, the flow of work was interrupted as participants were disturbed and puzzled by my presence. Using the body and bodily sensation to 'perceive' what is going on in the organization transforms the ethnographer from simply *doing* an ethnography (research*ing*) to doing research *with* the participants (Manning, 2016, 2018).

Ethnographic research sensitive to the body of the researcher interacting with other bodies in the research situation might even make use of a multi-sensory approach that includes sensory dimensions of aural, visual, olfactory, and haptic senses (Sparkes, 2009). Through self-reflexivity drawing on a multi-sensory approach, the researcher becomes aware of how her own ways of being, relating, and acting influence the research situation. Being attentive to all sensory feelings rushing through the body and mind will help the researcher to examine and unsettle the key assumptions consciously or subconsciously guiding her research (Cunliffe & Karunanayake, 2013). All these bodily sensations of discomfort and misattunement are 'part and parcel' of the data generated. To be sure, my bodily sensations while doing fieldwork are not particularly extraordinary nor generalizable as bodily sensations and relations will vary according to the social setting of fieldwork. But my focus on the body points to how my engaged ethnographic methods filled me with bodily sensations that, in turn, affect my interpretation of data.

Second, a focus on bodies in research can generate more sensitivity to the study of working bodies, to try to grasp the sensory aspects of the working practice. Using the body as the subject of perception, we might be able to grasp how we are "in the work through the body, just as we perceive the work with our body" (Merleau-Ponty & Smith, 1962, p. 206). Such a perspective opens up research to questions such as how it actually feels like to work and how we use our bodily senses at the workplace interacting with artifacts and other working bodies. Considering how key sociological variables such as race, gender, age, and so on fundamentally influence the sensory component of worker's experiences, the bodily sensations of the working body are important to incorporate in future workplace research (Holck, 2018a).

3.5.2 *Method relevance to practice*

Reporting my results back to the organization was difficult. The HR department wanted to use my findings to help to increase team diversity in Clean. I described all the benefits of diversity in terms of dynamic and positive collaboration, well-being among team members, and how leading a diverse team strengthens leadership development. With my 'progressive agenda', I also presented all the factors hinder-ing the progressive impact of the difference in diverse teams: problems of ingroup-outgroup behavior; exclusion of members 'not fitting the norm'; the problem of white managers vs. brown janitors that reified 'inequality regime'; problems of miscommunications and lacking confidence in managers based on language prob-lems; and so on.

I took part in several confidential meetings and completed confidential reports to the HR department (related to the task of increasing team diversity). In such deliverables, I communicated my recommendation on how to make diversity and inclusion 'work' in teams: for instance, to upgrade Danish competencies among janitors and English competencies among managers in order to create a shared language that connects instead of disconnecting by creating misunderstandings – maybe even a bilingual policy; to nurture a greater sense of fairness and justice by ensuring more (demographic) representation at all levels in the organization; to create awareness around the use of stereotypes and the existence of ethnicized and gendered hierarchies hampering collaboration as well as the unfolding of indi-vidual skills and qualifications among employees; to invest time in nurturing social relations in teams to counter exclusive ingroup/outgroup behavior; to train leaders, who develop and empower employees and built relations instead of managers, who are monitoring work; to support leaders with knowledge on conflict resolution, intercultural competences/intelligence.

The HR officers would listen to my experiences and recommendations. Still, they longed for results on the positive impact of diversity to use for internal moti-vation to increase team diversity and externally brand the organization. They were reluctant to react to the more 'problematic' and less 'rosy' description of diversity. However, my research impacted the professional development of service managers and supervisors at the frontline level, as diversity management was made a part of their training. Furthermore, an inclusive talent program was initiated to recruit qualified (minority) frontline staff for positions higher up the organizational hierar-chy to (hopefully) alter the leadership composition in the long run.

My research may also hold relevance for leaders and HR consultants more broadly who wish to deploy and promote inclusive leadership practices. Paying attention to the body, bodily reactions, and the identity work that goes on in work-place relations might help to develop diversity-sensitive leadership. Indeed, at-tentiveness to the performativity and affects of bodies in social relations among leaders and employees can reveal the 'stickiness' of privilege and disadvantage, tying certain organizational groups together in 'communities of fate' based on their demographic characteristics (Ahmed, 2014; Holck, 2018a, 2018b). Pay attention to affect as relational, performative, sticky, and contagious – which are moods picked up by the body – can help leaders and HR consultants to focus on two im-portant considerations when dealing with diversity and inclusion. First, the leader is always affectively entangled in the social situation of interacting with employ-ees. Critically reflecting on this affective entanglement will help to recognize how affect makes it difficult to distinguish personal experiences from the supposedly actual, objective social interaction of working bodies. Attention to affect should help leaders critically examine their role in (re)producing sentiments and relations in social workplace situations.

Second, focusing on affective bodies and their interaction makes it possible to work with problems of inclusion and exclusion in groups and teams. I became aware of silos and inclusion/exclusion dynamics through scrutinizing my bodily reactions while in the fieldwork. To actively use the senses of the body in social interactions might even point to alternative ways of working to alter the workplace in favor of equity and fairness: for instance, by engaging in mixed-racial/gender, etc., in-group collaboration while subtly problematizing, disrupting, and challenging taken-for-granted practices of task and/or position distribution based on bodily characteristics (Holck, 2018a, 2018b). Awareness of bodily reactions and their internal 'contamination' might sensitize the leader to ways to promote confidence and trust, and lower anxiety in the face of change. Working from and through the body as a leader might enable a more engaged, compassionate, resistant, and pluralistic workplace ethics to counter strong organizational tendencies toward control, homogeneity, discrimination, and domination (Pullen & Rhodes, 2015).

3.6 Concluding discussion

Affective workplace ethnography is grounded in post-qualitative methodologies and theorized as a type of research that acknowledges various data-driven elements such as involved actors, accessed texts, and language (Gherardi, 2019; Holck, 2018a). As such, it is a vital attribution to a researcher's scholarly approach, especially when exploring a wider angle of a particular subject within the workplace context. Even more so, as the diversity in and around organizations increases, workplace ethnographic methods become progressively relevant as they make the researcher sensitive to the delicate, affective, complex, and subtle aspects of diverse workplaces and work populations.

Through the lens of affect and bodies at work, this chapter highlights how workplace ethnographic fieldwork is indeed influenced by the body of the researcher and the interaction with other bodies in the fieldwork setting. I explore how workplace ethnographic research might be developed into a self-reflexive and multi-sensory approach to make the researcher attentive to how her body's relating to and interacting with participants influence the research situation. Being attentive to all the sensory feelings rushing through the body and mind will help the researcher to examine and unsettle the key assumptions consciously or subconsciously guiding research. Focusing on research instances of affect as "the personal, political and ethical considerations of our research … can result in more transparent and informed research accounts that recognize the lived experiences of those who have been marginalized in mainstream academic discourse" (Manning, 2018, p. 312). Drawing on the author's bodily sensations while doing ethnographic research can thus help to bring along a more reflexive, holistic, and context-sensitive research.

These implications reach beyond academia as leaders and HR managers should be aware of the implications of the body while dealing with a diverse organizational context. Leaders and HR consultants' sensitivity to the moods created by the interacting bodies might even be used as a lever to create a more just and equal workplace: Paying attention to affective shifts, leaders hold the possibility to use these instances to "destabili[se] categories and categorizations which may otherwise go unquestioned", as proposed by Kennedy-Macfoy and Nielsen (2012, p. 145).

However, there are certain limitations to doing longitudinal workplace ethnographic research: It is high time and resource-consuming for both researcher and the research site. Furthermore, more and more research is being conducted online, for instance, using Teams or Zoom to do individual and focal group interviews, meetings, training, and presentations as online interaction with research sites. Online interaction has facilitated easier access and faster generation of data, and research sites tend to wish for a mix of both physical and online interaction even beyond the peak of the Covid-19 pandemic. A limited physical presence while doing fieldwork makes it more difficult – but not impossible, however – to dwell on bodies and bodily relations at the workplace.

This chapter does not advocate a centricity of the body only but emphasizes the omission or overlooking of the body when conducting workplace studies or practicing (inclusive) leadership and diversity management. However, attention to the moods and the sensations of the body should not paralyze or overwhelm researchers and leaders or restrain them from taking painful actions 'stirring up the mood' of the workplace. Indeed, stirring up the mood through changing bodily interactions and relations may be a way to alter the unequal distribution of resources of power and influence, but such change attempts should be deployed with caution and awareness of the consequences for future workplace relations.

3.7   Further Reading

The notion of affective ethnography (Gherardi, 2019; Holck, 2018a) may act as a lever to orient yourself in relation of the body and bodily interaction while in the field. Affective ethnography is a research practice that acknowledges that all elements – texts, actors, materialities, language, and agencies – are already entangled in complex ways. Researchers, who wish to embark on affective ethnographic work can use both of these articles for inspiration and to further develop the field. To combine affective ethnographic research with a focus on the body, Merleau-Ponty and Smith (1962; see also Mikkelsen, 2022; Shilling, 2012), is a great way to be introduced to the body in research. Furthermore, Longhurst et al. (2008) offer a great way to explore how the body can act as a tool through which all interactions and emotions are filtered when accessing research subjects and their geographies.

Finally, if you want to explore the aspect of self-reflexivity and positionality in ethnographic research, the articles by Manning (2016, 2018) discuss how management scholars might (re)negotiate the complexities of positionality and representation, and how this approach helped her to become a decolonized feminist researcher. She claims that deploying a decolonial feminist approach to ethnography enabled her to identify positionality and representation as the critical complexities of engaging in research with marginalized 'others', and at the same time, how dealing with these aspects of research helped her to identify tools to address the complexities of positionality and representation.

Acknowledgments

The author would like to acknowledge the financial support of the Innovation Foundation (Denmark). Thank you to Florence Villesèche for her input and support in finalizing this chapter.

Notes

1 The name of the participants and of the organization have been changed to protect their identities.
2 Clean has their own definition of team diversity, which is max. 70% homogeneity in terms of gender, generation, and racial background of team members, which guided my definition of what characterized a diverse team.

References

Acker, J. (2011). Inequality regimes. In J. Z. Spade & C. G. Valentine (Eds.), *The kaleidoscope of gender: Prisms, patterns, and possibilities* (pp. 355–365). Pine Forge.

Ahmed, S. (2000). *Strange encounters: Embodied others in post-coloniality*. Routledge.

Ahmed, S. (2004). Declarations of whiteness: The non-performativity of anti-racism. *Borderlands E-Journal, 3*(2). http://www.borderlands.net.au/vol3no2_2004/ahmed_ declarations.htm

Ahmed, S. (2007). A phenomenology of whiteness. *Feminist Theory, 8*(2), 149–168.

Ahmed, S. (2014). Not in the mood. *New Formations, 82*(82), 13–28.

Ahmed, S. (2017). *Living a feminist life*. Duke University Press.

Andreassen, R., & Myong, L. (2017). Race, gender, and researcher positionality analysed through memory work. *Nordic Journal of Migration Research, 7*(2), 97–104.

Ashcraft, K. L. (2013). The glass slipper: "Incorporating" occupational identity in management studies. *Academy of Management Review, 38*(1), 6–31.

Ashcraft, K. L. (2018). Critical complicity: The feel of difference at work in home and field. *Management Learning, 49*(5), 613–623.

Clifford, J. (1983). On ethnographic authority. *Representations, 2*(Spring), 118–246.

Cunliffe, A. L. (2010). Retelling tales of the field: In search of organizational ethnography 20 years on. *Organizational Research Methods, 13*(2), 224–239.

Cunliffe, A. L., & Karunanayake, G. (2013). Working within hyphen-spaces in ethnographic research: Implications for research identities and practice. *Organizational Research Methods, 16*(3), 364–392.

Dar, S. (2018). Decolonizing the boundary-object. *Organization Studies, 39*(4), 565–584.

Flores, G. M. (2015). Discovering a hidden privilege: Ethnography in multiracial organizations as an outsider within. *Ethnography, 17*(2), 190–212.

Gherardi, S. (2019). Theorizing affective ethnography for organization studies. *Organization, 26*(6), 741–760.

Gilmore, S., & Kenny, K. (2015). Work-worlds colliding: Self-reflexivity, power and emotion in organizational ethnography. *Human Relations, 68*(1), 55–78.

Hibbert, P., Sillince, J., Diefenbach, T., & Cunliffe, A. L. (2014). Relationally reflexive practice: A generative approach to theory development in qualitative research. *Organizational Research Methods, 17*(3), 278–298.

Holck, L. (2016). Putting diversity to work: An empirical Analysis of how change efforts targeting organizational inequality failed. *Equality, Diversity and Inclusion: An International Journal, 35*(4), 296–307.

Holck, L. (2018a). Affective ethnography: Reflections on the application of "useful" research on workplace diversity. *Qualitative Research in Organizations and Management: An International Journal, 13*(3), 218–234.

Holck, L. (2018b). Unequal by structure: Exploring the structural embeddedness of organizational diversity. *Organization, 25*(2), 242–259.

Holck, L., & Muhr, S. L. (2020). White bodies in postcolonial ethnographic research. In S. N. Just, A. Risberg, & F. Villesèche (Eds.), *The Routledge companion to organizational diversity research methods* (pp. 36–49). Routledge.

Kennedy-Macfoy, M., & Nielsen, H. P. (2012). We need to talk about what race feels like! Using memory work to analyse the production of race and ethnicity in research encounters. *Qualitative Studies, 3*(2), 133–149.

Longhurst, R., Ho, E., & Johnston, L. (2008). Using 'the body' as an 'instrument of research': kimch'i and pavlova. *Area, 40*(2), 208–217.

Manning, J. (2016). Constructing a postcolonial feminist ethnography. *Journal of Organizational Ethnography, 5*(2), 90–105.

Manning, J. (2018). Becoming a decolonial feminist ethnographer: Addressing the complexities of positionality and representation. *Management Learning, 49*(3), 311–326.

McCorkel, J. A., & Myers, K. (2003). What difference does difference make? Position and privilege in the field. *Qualitative Sociology, 26*(2), 199–231.

Merleau-Ponty, M., & Smith, C. (1962). *Phenomenology of perception* (Vol. 26). Routledge.

Miettinen, R., Samra-Fredericks, D., & Yanow, D. (2009). Re-turn to practice: An introductory essay. *Organization Studies, 30*(12), 1309–1327.

Mikkelsen, E. N. (2022). Looking over your shoulder: Embodied responses to contamination in the emotional dirty work of prison officers. *Human Relations, 75*(9), 1770–1797.

Muhr, S. L., Holck, L., & Just, S. (2022). Ambiguous culture in Greenland police: Proposing a multi-dimensional framework of organizational culture for HRM theory and practice. *Human Resource Management Journal, 32*(4), 826–843. https://doi.org/10.1111/1748-8583.12472

Muhr, S. L., & Sullivan, K. R. (2013). "None so queer as folk": Gendered expectations and transgressive bodies in leadership. *Leadership, 9*(3), 416–435.

O'Reilly, K. (2012). Going native. In K. Reilly (Ed.) *Key concepts in ethnography*. Sage.

Pullen, A., & Rhodes, C. (2015). Ethics, embodiment and organizations. *Organization, 22*(2), 159–165.

Romani, L., Holck, L., & Risberg, A. (2019). Benevolent discrimination: Explaining how human resources professionals can be blind to the harm of diversity initiatives. *Organization, 26*(3), 371–390.

Silverman, D. (2001). *Interpreting qualitative data* (2nd ed.). Sage.

Shilling, C. (2012). The body and social theory. Sage.

Smit, B., & Onwuegbuzie, A. J. (2018). Observations in qualitative inquiry: When what you see is not what you see. *International Journal of Qualitative Methods, 17*(1), 1–3.

Sparkes, A. C. (2009). Ethnography and the senses: Challenges and possibilities. Qualitative Research in Sport and Exercise, 1(1), 21–35.

Van Maanen, J. (1991). The smile factory: Work at Disneyland. In P. J. Frost, L. F. Moore, M. R. Louis, C. C. Lundberg, & J. Martin (Eds.), *Reframing organizational culture* (pp. 58–76). Sage.

Van Maanen, J. (2011). Ethnography as work: Some rules of engagement. *Journal of Management Studies, 48*(1), 218–234.

Yanow, D. (2012). Organizational ethnography between toolbox and world-making. *Journal of Organizational Ethnography, 1*(1), 31–42.

Zanoni, P., & Janssens, M. (2015). The power of diversity discourses at work: On the interlocking nature of diversities and occupations. *Organization Studies, 36*(11), 1463–1483.

Zickar, M. J., & Carter, N. T. (2010). Reconnecting with the spirit of workplace ethnography: A historical review. *Organizational Research Methods, 13*(2), 304–319.

4 Digital ethnography

Understanding platform labour from within

Christian S. Ritter

Karlstad University, Sweden

4.1 Background

Long-term ethnographic fieldwork was established as a methodological approach in various social science disciplines early in the twentieth century. Early contributions to the ethnographic research method were made by the anthropologist Bronislaw Malinowski and the Chicago School of Sociology proponents Robert E. Park and Ernest Burgess, for example (Deegan, 2001). In the latter half of the twentieth century, ethnographic fieldwork was primarily conducted in human geography, sociology, anthropology, nursing, educational sciences, and, more recently, in business and management studies (e.g. Delamont & Atkinson, 2019). The rapid rise of digital media in the 1990s posed new challenges for ethnographic researchers. Committed to an immersive assessment of everyday routines evolving around Internet technologies (Coleman, 2010, p. 498), the scope of digital ethnographic research was initially circumscribed by the boundaries of online communities. In the wake of the cultural turn in communication research, numerous media scholars began to draw on ethnographic research to conduct in-depth investigations into digital media production, circulation, and consumption (Ardévol & Gómez-Cruz, 2012, p. 2). Dedicated to a practice-orientated approach (e.g. Couldry, 2004; Postill, 2010), digital ethnographers predominantly explore media practices within the multiple realms of localised lifeworlds, such as living rooms, pubs, parks, subway stations, and workplaces. Many contemporary professions are increasingly entangled in platformisation processes, creating forms of labour which are mainly oriented towards digital platforms. Ongoing platformisation processes pose a substantial challenge to contemporary qualitative inquiries. The platformisation of labour relates to the economic and infrastructural penetration of digital platforms into the workflows and taskscapes of employees (e.g. Nieborg & Poell, 2018). Ethnographic approaches are directed towards the local knowledge of a given community and the everyday practices of its members. By engaging with digital media technologies, digital ethnographers conduct iterative-inductive research that evolves throughout the investigation and draws on a family of methods (O'Reilly, 2005). The term digital primarily refers to all entities that can be reduced to binary code (Miller & Horst, 2012, p. 3), and present-day ethnographers increasingly incorporate digital methods into their research endeavours (e.g. Born &

DOI: 10.1201/9781003289845-4

Haworth, 2017). Therefore, digital ethnography can be seen as a strategy of inquiry involving iterative-inductive research, an immersion in the everyday practices of local communities, and a combination of traditional research techniques with digital methods. For instance, participant observation and in-depth interviews can be complemented with computational network analysis.

In this chapter, I draw on ethnographic research into a Norwegian software firm. During the course of this research project, I complemented participant observation in office spaces with the digital methods walkthrough and computational network analysis. Walkthroughs are widely understood as research techniques that systematically step through the various affordances of platforms. In contrast, computational network analysis entails network centrality measures and visualisations of relationships among platform users through data about platform practices, such as liking and retweeting. Present-day workplaces worldwide are increasingly entangled in complex taskscapes, which can be defined as arrays of related activities in professional settings (e.g. Ingold, 1993). The taskscapes of the researched software developers span across the physical settings of their office spaces and the digital platforms used at work. Indeed, the intertwinement of physical and digital contexts is inherent in many workplaces globally. Bridging the abstract division between physical and virtual units of analysis in workplace research, the ethnographic approach which I discuss in this chapter is anchored in an extended immersion in both the company's premises and platform interfaces. The main aim of this chapter is to demonstrate the substantial potential of ethnography for the study of professional groups in the digital economy. In the following section, I will consider methodological choices for researching the increasingly platform-orientated labour of knowledge workers. The next part of the chapter addresses the role of the walkthrough method in ethnographic research, followed by a computational analysis of affiliation networks within the researched software firm.

4.2 Argument: Demystifying the socio-technical assemblages of platform labour

For a long time, ethnographic workplace research was confined to a locale where professional activities were actually performed. However, in recent decades, the implementation of digital media technologies in numerous workflows has considerably transformed the post-Fordist office space, which posed numerous challenges to the practices of ethnographic workplace research and its traditional principles of holism and long-term immersion. A multitude of present-day professional groups inhabit screen-centred offices, and their skilled practices are mostly directed toward the interfaces of digital platforms. Previous generations of workplace researchers assessed knowledge practices, organisational discourses, and working orders among certain professional groups such as lawyers (Suchman, 2000). Numerous sociologists conducted ethnographic research to examine how digital media transformed organisational life, providing empirical studies about command and control centres, financial institutions, news media, and the construction industry

(Heath et al., 2000). After comparing different methods widely used in workplace research, Sellberg and Lindblom (2014) conclude that the methodological canon of the research area lacks in-depth analysis of workplace artefacts and their socio-historical development. To this day, the work practice studies of many management consultants are primarily based on etic frameworks, which examine corporate culture through a non-participatory lens and from an outsider's point of view. Hence, emic perspectives from within communities of practices are still much needed to broaden the scope of workplace studies (e.g. Jordan, 1996). Rooted in the local knowledge of employees, emic perspectives describe internal practices from within a community of practice and in terms that are meaningful to its members. Based on the case study of the Norwegian software firm, I suggest that the combination of ethnographic fieldwork and digital methods can bridge the hermeneutical gap between the subjective meanings employees attribute to their skilled practices and the meanings coded into platform affordances, allowing workplace researchers to design holistic projects for the study of Internet-saturated professions. Complementing onsite observations with digital methods, such as walkthroughs and computational network analysis, enables workplace researchers to put the rapid flow of platform practices on hold while unravelling the complex making of knowledge in the digital-physical continuum. By capturing the local meaning-making among employees and the connotations built into platform affordances, workplace researchers can critically evaluate the role of work performances and recommend corporate leaders to balance personal judgment and datafied assessment.

4.3   Example of use: Inside a software firm

Having studied the Norwegian company which develops business intelligence software for oil and gas corporations, this chapter assesses how digital methods can strengthen ethnographic research into platform labour. Today, an ever-increasing multitude of professions are entangled in platformisation processes, and the expansive variety of concomitant workflows involve tasks conducted on digital platforms, generating data traces for various forms of analytics. The research began at several industry events during which I became aware of the rising demand for software applications among oil and gas operators (Ritter, 2021). Such industry events were a great opportunity to arrange subsequent interviews with attendees at their companies' sites. Overall, I conducted fieldwork for 16 months at intersections between the Norwegian software industry and the global oil and gas industry. The purpose of the qualitative study was to trace the platform ecology and the practical expertise of software developers. After negotiating access to the firm, which is located in the town of Trondheim, I could carry out an onsite ethnography during a 3-month secondment. Conducting an ethnographic study on the premises of the Norwegian firm, I primarily examined how built-in platform metrics shaped the internal evaluation of work performances among software developers. By observing employees and participating in meetings on the firm's premises (see Figure 4.1), I could gain an in-depth understanding of the skilled practices required

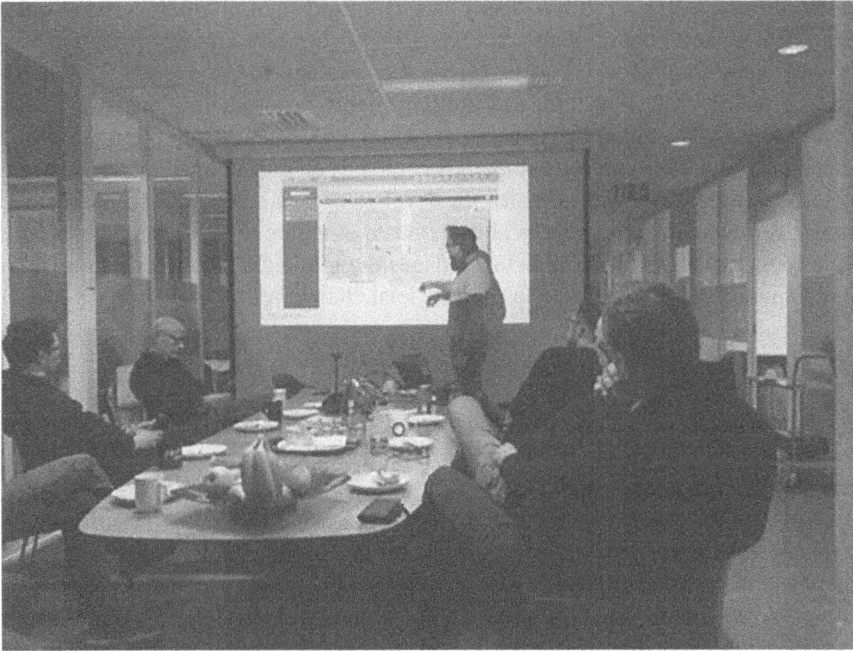

Figure 4.1 One of the meetings on the firm's premises (Photo by the author)

for the production of software. In addition, 30 in-depth interviews were conducted with software experts, including 18 from said firm (see Table 4.1).

In the autumn of 2018, I examined the production of business intelligence software for Norway's global oil and gas industry. Founded in 2001, the company initially aimed to sell expert reports on extraction equipment to oil and gas companies. At the time of this investigation, the company employed about 20 people. The firm's energy consultants conducted site visits to assess for instance drilling rigs. Based on their assessment, they compiled expert reports about operational risks. Soon after its inception, the firm expanded and created satellite offices in Qatar, UAE, and the USA, aiming to open its services to an increasingly globalised market in the oil and gas industry. Simultaneously, the firm established a cloud-based distribution system for its software products (Ritter, 2018, p. 2). When the

Table 4.1 Overview of research techniques

Method	Type of Evidence	Locations
Participant observation	Fieldnotes	Open-plan and single-person offices
In-depth interview	Interview transcripts	Single-person offices
Shadowing	Fieldnotes	Single-person offices
Walkthrough	Fieldnotes	Digital platforms

global oil price dropped dramatically in 2014, the firm's products were in much less demand, and consequently, the overseas offices were closed with some 60% of staff made redundant. The firm's response strategy to the economic crisis instigated a shift from developing consulting reports to producing business intelligence software for oil and gas operators (Ritter, 2019, p. 454). In the wake of this key managerial decision, the company's emphasis on consultants' expertise began to wane, whereas the role of software developers became central to its business model. Software development established itself as a core profession of the Internet era as developers build and maintain various digital media technologies that drive contemporary business operations and societal change.

4.3.1 Ethnography and interfaces

By discussing the challenges I faced while exploring the digitised taskscapes of professionals, I will demonstrate how the digital method walkthrough can be integrated into the ethnographic research process. After I was allocated a desk in the firm's main open-plan office, I began to conduct participant observation among the software developers, consultants, and sales personnel. While spending time with employees, I could attend meetings and shadow their work routines in the physical settings of the company site. Shadowing is an informal observational research technique for understanding how employees perform a specific task. A shadowing session can be seen as an observational tactic that allows ethnographers to elicit the subjective meanings people assign to their everyday practices. After the first few weeks of the secondment, I arranged shadowing sessions with members from each department. I asked my interlocutors during the shadowing sessions to explain their most common tasks. Having taken notes during these sessions, I conducted follow-up interviews with the participants (e.g. Ventura & Keinan-Guy, 2018, p. 5). The employees produced various narratives about their taskscapes during the shadowing sessions. In addition, the interview reinforced my understanding of their skilled practices, allowing me to cross-check central pieces of information. The narratives of interlocutors revealed how platforms or other artefacts were locally appropriated and how their hallmarks were categorised. The shadowing sessions and the accompanying in-depth interviews with software developers helped me comprehend their emic perspectives on their tasks, skilled practices, and internal relationships with colleagues and tools. Emic perspectives convey the local knowledge existing within a given community. Internal practices are described in terms that are meaningful to the members of a community of practice.

By attending onsite meetings, I could conduct further participant observations. During my stint at the company, I mainly participated in 'daily scrums' and 'all-hands', and such meetings helped me learn how employees made sense of their internal events, skilled practices, and digital platforms used to produce software. All-hands meetings were scheduled for Monday afternoons. The following revised fieldnote describes how this type of meeting unfolded in the researched company:

The meeting begins with a PowerPoint presentation by the consultant Emil. He elaborates on a planned update of one of the firm's software programs.

After various meetings with clients, Emil spent the last weeks designing the blueprint for a new analytics dashboard. He reports that the new design is in line with the requests from existing users. Based on their feedback and wishes, he presents a new tool to the Chief Executive Officer (CEO) Jan and the software developers who are supposed to write the code for the update in the upcoming weeks. The slides display the main elements of the dashboard. Emil's presentation sparks a tense debate between numerous consultants and software developers. The latter question repeatedly the feasibility of the calculations on which some features of the update are based. Jan requests technical information from both consultants and software developers. Finally, he adjourns the meeting prematurely. He suggests meeting an important client and user of the software before proceeding with the development of the update.

Immersing myself in the local meeting culture of the firm elucidated how employees interpreted the categories for professional roles and internal hierarchies. By writing fieldnotes, I documented the local knowledge of the employees. Acting as a participant-as-observer on the company site, I could, for instance, glean crucial information about the internal evaluations of work performances. Assessing skilled practices *in situ* also made possible the documentation of localised understandings of the project management framework scrum. Furthermore, onsite observations brought to light how energy consultants acquired skills from software developers. The underlying objective of such place-based observations was to elicit emic perspectives on internal practices, social relationships, and local artefacts.

Very soon into the secondment, I noticed that most employees spent their working days in front of a desktop and laptop computer screen. Traditional observational techniques could not provide any insights into how the affordances of platforms shaped the daily routines of software developers. Affordances can be seen as the material properties of digital objects that shape the ways in which they can be used. A few weeks into my placement, a leader of one of the software teams granted me access to the platforms GitHub, Slack, and Huboard and the software packages produced by the staff, which were distributed via the cloud. As the daily routines of software developers revolved around the interfaces of these internal platforms, the observational practices of workplace researchers require systematic documentation of platform interfaces. Such interfaces are primarily shared boundaries between computer systems, connecting hardware, software, and human users. Furthermore, they can be defined as mediation techniques and effects at the interstices of sociotechnical systems (Galloway, 2012, p. 33). By performing technical walkthroughs of such interfaces, I could gain a more comprehensive understanding of platform affordances' role in software production. The walkthrough method demonstrates how built-in mediators of platform interfaces construct symbolic meanings (Light et al., 2018). I documented the meanings assigned to the various affordances of interfaces, including the design and layout of their menus, icons, and buttons. At one software meeting, I learned that a client had reported problems with an access key to her software package during a support call. Mark, the software team leader, allocated the task of repairing the access portal to two team members on the GitHub platform, a hosting service for software development and version control. The following extract from a revised observation record contains a technical walk-through of a central GitHub interface:

> The web browser displays a menu bar on the top of a code repository. The menu bar contains the buttons 'code', 'issues', 'pull request', 'projects', 'wiki' and 'insights'. Under 'issues' a long list of items can be found. Issues are features of GitHub that allow its users to track their work. The progress of an issue is shown on a timeline. Mark creates an issue for the access key problem. He titled the issue 'Access bug' and entered a long description alongside a screenshot of an error message. He adds two members of the team to the issue. After the two team members complete their adjustments of the code, they attach the label 'PO test' to the issue. Subsequently, the prod-uct owner tests and approves the repaired feature of the software. The issue is forwarded to a 'UA test'. The user acceptance test is performed by a con-sultant who adds a comment about the user experience to the issue. Finally, Mark attaches the label 'Approved' to the issue. The various affordances of the code repository shape the ways in which the bug can be fixed and how employees report on their progress. The different buttons, labels, and menus on the platform GitHub structure the workflow for repairing a front-end bug.

During the ethnographic investigation into software making, the walkthrough method provided invaluable insights into how interface affordances invoked ways of organising tasks and collaborations. The underlying objective of this method was to demystify how platform interfaces construct ideologies of achievement and competitiveness. Technical walkthroughs enable researchers to critique the socio-cultural assumptions coded into platform affordances and thus govern the every-day practices of their users. The walkthrough method documents the symbolic meanings circulating through the interfaces of digital technologies while filling the lacunas left by participant observation directed towards places and events. The technical walkthroughs allowed me to trace how transformative mediators attribute meanings to the affordances of central platforms, such as Slack and GitHub.

The software firm under investigation mainly consisted of local places and inter-locking platforms where observational techniques could be employed. Combining participant observation and the walkthrough method enables workplace research-ers to trace the twofold construction of meaning which occurs in communities of practice where labour is directed towards digital platforms. The skilled practices in office spaces and on digital platforms were inseparable from the everyday lives of the firm's employees. However, the act of writing fieldnotes, which documented both local practices and platform practices, reveals the differences between the two observational techniques. The technical walkthroughs were conducted on my desk-top computer in an open-plan office. Whereas these fieldnotes primarily focused on the buttons, icons, and menus of interfaces, the fieldnotes based on participant observation described the local knowledge on the company premises. This type of fieldnote documented situations in the office space, including activities, actors, and small-scale events. While technical walkthroughs enable researchers to systematically trace the meanings generated by the affordances of interfaces, participant observation explicates how members of a given community of practice interpret their social relationships, skilled practices, and artefacts at work. Both types of fieldnotes complemented interview transcripts. The complete qualitative data set was analysed in accordance with the coding procedures of grounded theory, which ensures that theoretical claims are based on concrete evidence and inductive reasoning (e.g. Bowen, 2008; Jinghong et al., 2019). Combining the two observational methods initiates a dynamic interplay between centring and de-centring digital media. The technical walkthrough can be seen as a media-centric technique of observation. In contrast, participant observation in a local office enables researchers to foreground the ways in which employees engage in media practices. By exploring media *in situ*, researchers de-centre digital media. They acknowledge that media are inseparable from other technologies and activities through which they are experienced. Media are part of, and entangled with, broader settings and socio-technical relations within everyday worlds. Assessing everyday realities as embedded social practices, non-media-centric research elucidates the local contexts of digital media (e.g. Bräuchler & Budka, 2020; Moores, 2018). The tension between both data collection techniques makes possible a dialectical spiral that incrementally drives the research process.

4.3.2 Ethnography and computational network analysis

Although employing observational techniques on the premises of the researched firm provided great insights into employees' everyday routines, the dynamics of their digital labour are deeply embedded in often-obscured, access-restricted digital infrastructures. The software developers under investigation mainly used GitHub and Slack in their daily working lives. Ethnographic research into digital labour can integrate digital methods to gain a more comprehensive understanding of platform practices. Workplace researchers can conduct participant observation among software developers to elicit the specific meanings they attribute to their skilled practices. However, platform practices leave digital traces in backend databases; traces can be re-purposed for a critical digital analysis. Given this double character of platform practices, mixed-method approaches can illuminate the multiple facets of digital labour (e.g. Berthod et al., 2017). Digital methods research is based on natively digital data and requires medium-specific data collection tools. These medium-specific tools, such as the IssueCrawler and the Twitter Capture and Analysis Toolset, can substantially complement offline ethnographic research (Born & Haworth, 2017, p. 71). Computational network analysis is a digital method that can measure network centrality and provide visualisations of relationships within a given platform ecology (see Table 4.2). Network visualisations enable researchers to identify clusters, alliances, gatekeepers, liaisons, and bridges within a given network. Interpretations of network graphs can shed light on internal dynamics within professional groups. The underlying objective of employing computational

Table 4.2 Elements of computational network analysis

Method	Type of Evidence	Location	Data Collection Tool	Data Analysis Software
Computational network analysis	Network graph and centrality metrics	GitHub's API	DMI GitHubScraper	Gephi

network analysis on the production platform GitHub was to trace connections and collaborations among its users. The ethnographic descriptions of internal events and local places contextualise the outcomes of the computational network analysis. While ethnographic observations study the everyday practices and contexts surrounding platform practices, computational network analysis follows the medium and relies on natively digital data (e.g. Caliandro, 2018; Rogers, 2013).

Researchers conducting computational network analysis mostly retrieve relevant data from the Application Programming Interfaces (APIs) of digital platforms or develop their own scraping software to extract data from the front-end layer of websites. For the last decade, media researchers have regularly retrieved network data from the APIs of many popular platforms, including Facebook, Twitter, and YouTube. However, in the wake of the Facebook Cambridge Analytica affair, access to APIs of digital platforms was considerably restricted for academic researchers (e.g. Bruns, 2019; Puschmann, 2019). The toolbase of the research group Digital Methods Initiative (DMI), which is located in Amsterdam, provides several data retrieval tools for the digital platform GitHub. In order to collect network data for this particular case study, I used the tool GitHubScraper. As a GitHub user with invited access to the firm's private repositories, I could use my login credentials for GitHub to retrieve network data about them using the GitHubScraper. In contrast to the publicly available code repositories for open-access software, firms producing proprietary software store the source code of their products in private repositories, which only employees or partner companies can access. In these digital locations, software developers can control the versions and updates of specific software packages. They also uploaded minutes from developer meetings, and documented the history of a software program in release notes on GitHub. Thanks to my GitHub login credentials, I was able to generate an access token for the GitHub API. Since I had regularly observed the platform practices of the firm's software developers on GitHub, I knew their GitHub usernames. I could compile of list of users from all the private repositories the firm maintained on GitHub. I used this list as a query in the GitHubScraper to assemble a network data set about the private repositories of the researched firm. The retrieved GEXF file could be opened in network visualisation software, such as Gephi and visone. The data set was comprised of 215 network nodes and 238 network edges. The bipartite network contained two sets of nodes, namely repositories and users. The edges of the network relate to the platform practice 'watching' and are based on the digital traces that platform users leave while engaging with the affordances of the platform. The platform practice 'watching' on GitHub is similar to 'following' or 'subscribing' on other platforms.

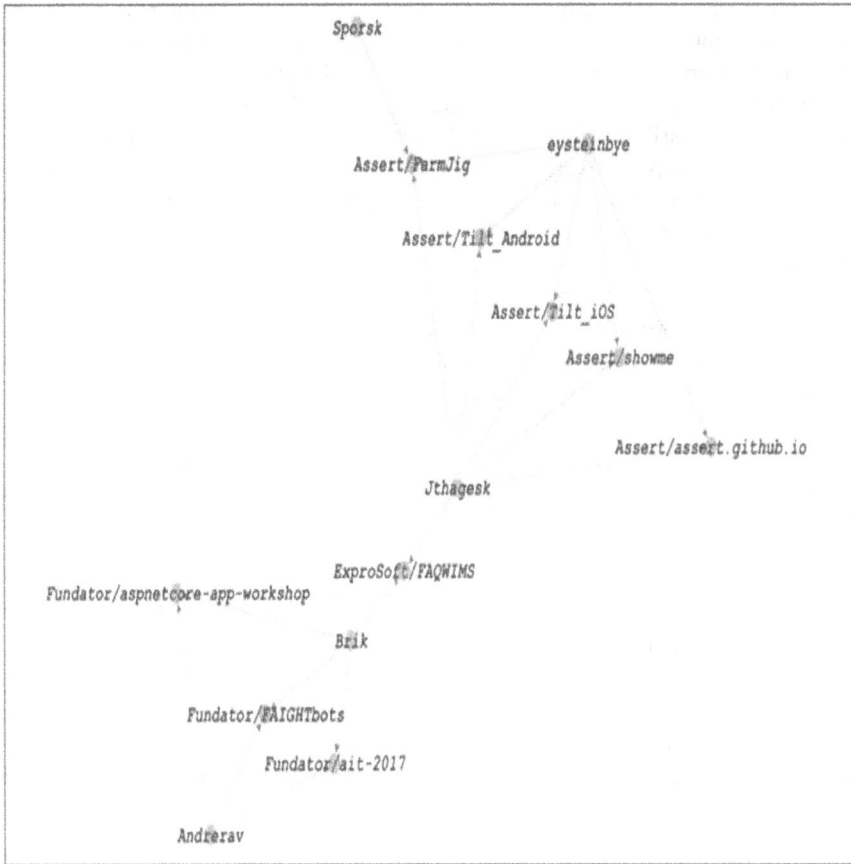

Figure 4.2 Bipartite network of private GitHub repositories

Software developers 'watch' code repositories relevant to their work and interests, reflecting their expertise in code libraries.

The affiliation network in Figure 4.2 shows the GitHub usernames of software developers working in the researched firm and the names of code repositories. The edges of the directed network point from software developers to code repositories, which is symbolised by a small arrow. The network visualisation was generated with the layout algorithm Force Atlas 2 and filtered to a degree ≥ 2 (Jacomy et al., 2014). The graph displays a sub-network made up of two clusters. The repository between the two clusters (ExproSoft/FAQWIMS) can be interpreted as a liaison through which the communication among the software developers flows. This repository contains basic information about the software product Well Integrity Management Systems (WIMS).

As I learned during the fieldwork on the company's premises, the software developer Jthagesk contributed to the liaising repository and further repositories of the researched firm. The software developer Brik was hired by the researched firm just weeks before I started my secondment. Prior to joining the researched software firm, she was employed by another local software firm whose repositories are displayed in the cluster on the left-hand side of Figure 4.2. In terms of the repositories of her new firm, Brik is only connected to the liaising one. Positioned between her current and previous company, she is the only person connected to the repositories of both. Thus, she has a significant role in sharing information on recent developments. The visualisation of the affiliation network indicates which repositories the software developers belong to, thereby providing insights into their professional trajectories and collaborations. Computational network analysis can provide visual evidence for the professional relationships among employees on a given digital platform. Information obtained during the onsite ethnography informed the interpretation of network graphs. The local knowledge documented during participant observation enhanced my understanding of specific nodes and edges. Integrating ethnographic fieldwork with computational network analysis also advanced my knowledge of how employees acquired and shared digital skills in the studied organisation.

4.4   Implications: Doing digital ethnography in the workplace

This case study illustrates how onsite ethnographic research and digital methods can be integrated, which has various implications for workplace researchers and practitioners. The qualitative method walkthrough illuminates how non-human actants, such as icons, menus, and buttons, shape the working lives of professionals whose labour is orientated towards digital platforms. In contrast, computational network analysis is a quantitative research technique which allows researchers to calculate network centrality scores for specific nodes in a given network and to visualise clusters and relationships in network graphs. Workplace studies can benefit in multiple ways from ethnographic research that complements place-based observations with digital methods. This ethnographic approach to workplaces can elicit the meanings that employees assign to their skilled practices and trace the meanings built into platform affordances. In addition to the methods walkthrough and computational network analysis, ethnographers can incorporate further digital methods, including the YouTube data tools, which can provide insights into channel and video networks, and the visualisation software IssueCrawler, which performs, among other things, a co-link analysis of Uniform Resource Locators (URLs). Fieldwork on digital platforms and production sites can, in turn, reduce the dependency of Internet researchers on standard API data (Venturini & Rogers, 2019, p. 537).

4.4.1 Research

Ethnographic researchers whose observations are directed towards platform interfaces encounter various theoretical challenges throughout the data collection and analysis procedures. In this chapter, I make a case for overcoming such a conundrum through the walkthrough method and computational network analysis.

The walkthrough method posits the substantial role of non-human actants in platform ecologies (Light et al., 2018). Ethnographic workplace researchers seek to holistically explore the interlocking worlds of digital platforms and office compounds. The ethnographic case of software making in Trondheim indicates that the taskscapes of software developers are located within complex socio-technical assemblages. Ethnographic research can shed light on the various components of assemblages holding together software developers, product owners, consultants, salespeople, and software users. Although cloud-based software has widely been celebrated as a free-flowing technology, its production, distribution, and usage are deeply grounded in material infrastructures, local hierarchies, and transnational dependencies. Digital ethnography can elucidate such far-reaching entanglements and identify locations for human agency in the increasingly platformised working life. While contemporary Internet technologies mostly involve automation processes and machine-learning algorithms, digital ethnographers have the ability to increase the transparency of the opaque architecture of platforms. Researchers can gain an in-depth understanding of contemporary workplaces by assessing local practices in workplaces and mediation practices in interfaces. Ethnography reveals how employees both engage with and resist the ongoing platformisation of digital labour. The discussed ethnographic approach to workplaces enables researchers to trace professional hierarchies, performance appraisals, and power relations among intermediaries within the socio-technical assemblages of the digital economy. Workplace scholars from different disciplinary backgrounds can thoroughly examine the role of digital platforms, such as Slack, Teams, and Jammer. By employing the localised walkthrough to understand the affordances of digital platforms, digital ethnographers can document hierarchies and divisions of labour within companies. Although computational network analysis may involve the constraints placed on data collection, this digital research technique can provide crucial insights into platform practices. Network visualisations can help improve communication channels within departments and among branches of a given corporation. Future ethnographic investigations can use the following or related research questions as points of departure:

- In what ways do digital platforms reconfigure company hierarchies?
- How do digital technologies facilitate creative collaborations among employees?
- Which non-human elements reshape contemporary workplaces?
- How are gender, race, age, ethnicity, ability, and sexualities negotiated in the day-to-day life of employees?
- To what extent are quantification technologies used in evaluations of work performance?

4.4.2 Practice

Practitioners can use the research techniques of digital ethnography to examine the actual uses of internal platforms and the well-being of employees. In addition to employing surveys and experiments, workplace researchers can combine participant observations and digital methods to inform the decision-making of trade unions, human resource departments, and consulting firms. These stakeholders can benefit from ethnographic research since they gain first-hand insights into how employees interpret their everyday practices in digitised taskscapes. In addition, different digital technologies implementation phases can be thoroughly assessed. The polyvocal accounts that ethnographic research provides can enhance employee satisfaction and democratic decision-making in firms. Based on ethnographic materials, employees can develop user guides for internal technologies, from which new staff can significantly benefit. And finally, ethnographic research into workplaces can also inform managerial reactions to ongoing digital disruptions and shape decision-making on purchasing innovative digital technologies.

4.5   Conclusions

Based on an in-depth investigation into a Norwegian software firm, this chapter explored how digital methods can strengthen ethnographic fieldwork. The proposed ethnographic approach complemented participant observation in the physical settings of open-plan offices with technical walkthroughs of platform interfaces and computational network analysis. The main aim of this chapter was to demonstrate the substantial potential of ethnography to enhance the study of professional groups in the digital economy. Technical walkthroughs allow workplace researchers to systematically document how platform interfaces reconfigure present-day work environments. In contrast, computational network analysis can provide invaluable insights into internal collaborations and connections among employees and firms. Integrating ethnographic research and digital methods advances the knowledge of the contemporary workplace by considering the medium-specific properties of digital platforms. Based on this case study, I suggest that combining ethnographic fieldwork with digital methods can bridge the hermeneutical gap between employees' subjective meanings of their skilled practices and the meanings coded into platform affordances, allowing workplace researchers to design holistic projects for the study of Internet-saturated professions.

Ethnographic fieldwork anchored in the interfaces of platforms is particularly suited to researching the socio-technical dynamics of numerous professional groups within the global economy. This approach is pertinent for exploratory assessments of the implementation of digital technologies in the workplaces of knowledge workers. Digital ethnographers can, for instance, illuminate current transformations in the financial sector, the influencer industry, the education sector, the gaming industry, the health sector, and co-working spaces. Furthermore, in-depth ethnographic studies can contribute to discourses on diversity and gender in the workplace and trace knowledge transfers among professional groups. Finally, workplace researchers can explore the political economy of transnational production networks, the situated construction of professional identities, and emerging forms of digital enskilment.

Nonetheless, this investigation into software developers in Norway also revealed several shortcomings of ethnographic tactics. A major limitation of ethnographic research on company premises concerns the restricted access of researchers to high-level meetings, documents, backend interfaces, and digital data. Ethnographic immersions in company sites remain partial, and work-based researchers are requested to sign non-disclosure agreements about the intellectual property of the researched firms. While ethnographic research provides great insights into the experience and attitudes of employees, this branch of qualitative research can be a very time-consuming endeavour. Researchers may require job-specific training and long-term stays at companies' sites to produce substantial results. The exploratory nature of ethnographic research regularly enables researchers to discover new trends and developments from within an organisation. However, employees might not be willing to discuss workplace problems due to hierarchical company structures. The outcomes from the ethnographic investigation into software making and the accompanying methodological reflections underscore that workplace researchers should establish, and adhere to, high ethical standards to protect employees' identities and interests. Future ethnographic investigations into workplace phenomena can assess the appropriateness of further digital tools for ethnographic research, and experiment with new forms of collaboration between researchers and enterprises.

4.6 Further Reading

Beuving, J. (2020). Ethnography's future in the age of big data. *Information, Communication, and Society, 23*(11), 1625–1639. https://doi.org/10.1080/1369118X.2019.1602664

Discussing the claims made in the best-selling book *Everybody Lies*, Beuving's essay opposes the different epistemological terminologies used in the ethnographic tradition and in big data research. Furthermore, avenues for combining ethnography and big data research are provided.

Chee, F. (2015). Online games and digital ethnography. In R. Mansell & P. Ang (Eds.), *International encyclopedia of digital communication and society* (pp. 1–6). Wiley.

Following a brief overview of the history of ethnography, the chapter proposes ethnographic approaches to games and gamers. Chee provides a comprehensive introduction to ethnographic research into gaming cultures. Since games are increasingly considered extensions of everyday life, video and smartphone games have become central places for interaction and identity formation.

de Seta, G. (2020). Three lies of digital ethnography. *Journal of Digital Social Research, 2*(1), 77–97. https://doi.org/10.33621/jdsr.v2i1.24

Updating Gary Alan Fine's piece on the *Ten Lies of Ethnography*, de Seta identifies three lies of digital ethnography: the networked field-weaver, the eager participant-lurker, and the expert fabricator. Grounded in ethnographic research into digital media use in China, the article stresses the need to confront methodological illusions.

Ritter, C. (2022). Unlocking heritage in situ: Tourist places and augmented reality in Estonia. In E. Costa, P. Lange, N. Haynes & J. Sinanan (Eds.), *The Routledge companion to media anthropology*. Routledge.

Based on long-term fieldwork in Estonia, the chapter assesses how augmented reality technology reorders tourist imaginaries about Estonia. Analysing the design aspirations and visons of Estonian-based app makers, the investigation demonstrates how quantification technologies shaped their app production.

References

Ardévol, E., & E. Gómez-Cruz. (2012). Digital ethnography and media practices. In V. Mayer, & J. Nerone (Eds.), *The international encyclopedia of media studies* (pp. 498–518). Wiley.

Berthod, O., Grothe-Hammer, M., & Sydow, J. (2017). Network ethnography: A mixed-method approach for the study of practices in interorganizational settings. *Organizational Research Methods, 20*(2), 299–323. https://doi.org/10.1177/1094428116633872

Born, G., & Haworth, C. (2017). Mixing it: Digital ethnography and online research methods - A tale of two global digital music genres. In L. Hjorth, H. Horst, A. Galloway, & G. Bell (Eds.), *The Routledge companion to digital ethnography* (pp. 96–112). Routledge.

Bowen, G. (2008). Grounded theory and sensitizing concepts. *International Journal of Qualitative Methods, 5*(3), 12–23. https://doi.org/10.1177/160940690600500304

Bräuchler, B., & Budka, P. (2020). Introduction. In P. Budka & B. Bräuchler (Eds.), *Theorising media and conflict* (pp. 3–30). Berghahn Books.

Bruns, A. (2019). After the 'APIcalypse': Social media platforms and their fight against critical scholarly research. *Information, Communication & Society, 22*(11), 1544–1566. https://doi.org/10.1080/1369118X.2019.1637447

Caliandro, A. (2018). Digital methods for ethnography: Analytical concepts for ethnographers exploring social media environments. *Journal of Contemporary Ethnography, 47*(5), 551–578. https://doi.org/10.1177/0891241617702960

Coleman, G. (2010). Ethnographic approaches to digital media. *Annual Review of Anthropology, 39*(1), 487–505. https://doi.org/10.1146/annurev.anthro.012809.104945

Couldry, N. (2004). Theorising media as practice. *Social Semiotics, 14*(2), 115–132. https://doi.org/10.1080/1035033042000238295

Deegan, M. (2001). The Chicago school of ethnography. In P. Atkinson, A. Coffey, S. Delamont, J. Lofland, & L. Lofland (Eds.), *Handbook of ethnography* (pp. 11–23). Sage. https://dx.doi.org/10.4135/9781848608337

Delamont, S., & Atkinson, P. (2019). Ethnographic fieldwork. In P. Atkinson, S. Delamont, A. Cernat, J.W. Sakshaug, & R. Williams (Eds.), *Sage research methods foundations*. Sage. https://dx.doi.org/10.4135/9781526421036771315

Galloway, A. (2012). *The interface effect*. Polity Press.

Heath, C., Knoblauch, H., & Luff, P. (2000). Technology and social interaction: The emergence of 'workplace studies'. *The British Journal of Sociology, 51*(2), 299–320. https://doi.org/10.1111/j.1468-4446.2000.00299.x

Ingold, T. (1993). The temporality of the landscape. *World Archaeology, 25*(2), 152–174. https://www.jstor.org/stable/124811

Jacomy, M., Venturini, T., Heymann, S., & Bastian, M. (2014). ForceAtlas2, a continuous graph layout algorithm for handy network visualization designed for the Gephi software. *PLoS One, 9*(6), 1–12. https://doi.org/10.1371/journal.pone.0098679

Jinghong, X., Xinyang, Y., Shiming, H., & Wenbing, C. (2019). Grounded theory in journalism and communication studies in the Chinese mainland (2004–2017): Status quo and problems. *Global Media and China, 4*(1), 138–152. https://doi.org/10.1177/2059436418821043

Jordan, B. (1996). Chapter 3 ethnographic workplace studies and computer supported cooperative work. In D. Shapiro, M. Tauber, & R. Traunmüller (Eds.), *The design of computer-supported cooperative work and groupware systems* (pp. 17–42). Elsevier Science. https://doi.org/10.1016/S0923-8433(96)80005-0

Light, B., Burgess, J., & Duguay, S. (2018). The walkthrough method: An approach to the study of apps. *New Media & Society, 20*(3), 881–900. https://doi.org/10.1177/1461444816675438

Miller, D., & Horst, H. (2012). The digital and the human: A prospectus for digital anthropology. In H. Horst & D. Miller (Eds.), *Digital anthropology* (pp. 3–38). Berghahn Books.

Moores, S. (2018). *Digital orientations: Non-media-centric media studies and nonrepresentational theories of practices*. Peter Lang.

Nieborg, D., & Poell, T. (2018). The platformization of cultural production: Theorizing the contingent cultural commodity. *New Media & Society, 20*(11), 4275–4292. https://doi.org/10.1177/1461444818769694

O'Reilly, K. (2005). *Ethnographic methods*. Routledge.

Postill, J. (2010). Introduction: Theorising media and practice. In B. Bräuchler & J. Postill (Eds.), *Theorising media and practice* (pp. 1–32). Berghahn Books.

Puschmann, C. (2019). An end to the wild west of social media research: A response to Axel Bruns. *Information, Communication & Society, 22*(11), 1582–1589. https://doi.org/10.1080/1369118X.2019.1646300

Ritter, C. (2018). Distributing services through the cloud infrastructure: The case of a Norwegian software firm. *AoIR Selected Papers of Internet Research, 8*, 1–4. https://doi.org/10.5210/spir.v2018i0.10503

Ritter, C. (2019). Shifting needs for expertise: Digitized taskscapes in the Norwegian oil and gas industry. *The Extractive Industries and Society, 6*(2), 454–462. https://doi.org/10.1016/j.exis.2019.03.006

Ritter, C. (2021). Rethinking digital ethnography: A qualitative approach to understanding interfaces. *Qualitative Research, 22*(6), 1–17. https://doi.org/10.1177/14687941211000540

Rogers, R. (2013). *Digital methods*. MIT Press.

Sellberg, C., & Lindblom, J. (2014). Comparing methods for workplace studies: A theoretical and empirical analysis. *Cognition, Technology & Work, 16*, 467–486. https://doi.org/10.1007/s10111-014-0273-3

Suchman, L. (2000). Making a case: 'Knowledge' and 'routine' work in document production. In P. Luff, J. Hindmarsh, & C. Heath (Eds.), *Workplace studies: Recovering work practice and informing system design* (pp. 29–45). Cambridge University Press.

Ventura, J., & Keinan-Guy, I. (2018). Shadowing as a central research method in medical design anthropology. In *SAGE Research Methods Cases*. Sage.

Venturini, T., & Rogers, R. (2019). 'API-based research' or how can digital sociology and journalism studies learn from the Facebook and Cambridge Analytica data breach. *Digital Journalism, 7*(4), 532–540. https://doi.org/10.1080/21670811.2019.1591927

5 Critical discourse analysis

Studying the symbolic aspects of workplaces

Masoud Shadnam

Sharif University of Technology, Iran

5.1 Background

Over the past three decades, workplace research has become the host of a growing wave of studies focusing on workplace semiotic aspects. Among the approaches that are used in this wave, Critical Discourse Analysis (CDA) has been a prominent one and is increasingly employed in workplace research. In this section, I introduce CDA by reviewing its history of emergence in linguistics and then its appropriation in workplace studies. The historical perspective enables us to highlight the situated character of CDA and its specificities as a particular type of scholarly work (Hodgson, 2001; Jovanović, 2011).

The birth of CDA in its current form was in the 1970s among European researchers in the discipline of linguistics (Wodak, 2001). This historical background points toward two fundamental characteristics: First, CDA was developed among those for whom the customary object of study was language. So as the banner "discourse analysis" also suggests language is at the core of CDA and research based on CDA focuses on those topics and issues in which linguistic aspects are salient. But it is important to note that up until the 1970s, linguistics was largely concentrated on studying the formal aspects of language (Chomsky, 1957), and even the newly created field of sociolinguistics was mainly aimed at examining the effects of different aspects of society on language and its variations in different contexts (Bernstein, 1973a, 1973b, 1977; Labov, 1966, 1972). One of the points where CDA departed from the mainstream of linguistics and sociolinguistics at the time was through its explicit focus on the impact of language on various social problems and their underlying societal structures. For CDA scholars, studying language had to move beyond understanding its internal structure or examining it as a dependent variable toward studying language as constitutive of social phenomena. Through language, people communicate with one another about the world around them and define what is real or unreal, natural or weird, moral or immoral (Shadnam, 2020). For example, Fairclough used CDA to illustrate how global capitalism in its current neo-liberal form "is pervasively constructed as external, unchangeable, and unquestionable – the simple 'fact of life' which we must respond to" (2001, p. 129).

The second fundamental characteristic of CDA is rooted in the geography of its birth. The influence of the European intellectual atmosphere on CDA is clear

DOI: 10.1201/9781003289845-5

in its grounding in continental philosophy and, in particular, in its critical approach toward linguistic research. The term "critical" in critical discourse analysis refers to the view that science always serves a purpose and the purpose of CDA is to reveal and undo injustice in its various forms. So unlike other strands of scientific work that contemplate and reproduce what is already accepted, CDA is a practice for addressing social problems (Fairclough et al., 2011). However, many serious social problems are rooted in ideologies, i.e., deep-seated and organized realms of discursive meaning and practice through which people feel, reason, desire, and imagine (Eagleton, 1991). Thus, the aim of CDA as a critical scientific practice is to demystify these ideologies and help the oppressed people toward emancipation from the associated problems (Wodak, 2001). Accordingly, the starting point of a CDA study, as discussed in more detail below, is a serious social problem with a discursive dimension. For example, in one of the early works in the current form of CDA studies, van Dijk (1984) focused on the problem of ethnic prejudice in the Netherlands and analyzed the discourse characteristics of prejudiced talk to understand the processes in which racist beliefs and attitudes are formed and diffused.

Within the realm of workplace research, scholars gradually recognized the merits of CDA for addressing the symbolic aspects of workplaces, which led to the early wave of workplace research drawing on CDA (Hardy & Phillips, 1999; Hardy et al., 2000; Phillips & Hardy, 1997). Taking a social constructionist perspective (Berger & Luckmann, 1966) and loosely drawing on the works of discourse analysts (Fairclough, 1992; Parker, 1992), this wave of work introduced a simple set of concepts to describe how discourse shapes social reality. In this conception, social reality is structured mainly through the discursive construction of three entities: concepts, objects, and subject positions. The first one, "concepts", refers to ideas and theories through which we understand the world around us and communicate with one another. For example, "health" is a concept, i.e., a historically and culturally contingent construction produced by the ongoing discursive practices of social actors (Turner, 2000). Although concepts are more or less contested and variable over time, they carry a sense of "rightness". So, for instance, when people use the concept of health in their daily discourse, they normally view it as the right descriptor of the outside reality rather than a constructed idea (Hardy et al., 2000; Phillips & Hardy, 1997). The second discursive entity, "objects", refers to concepts attached to a material referent. People construct objects when they employ concepts to make sense of the world around them (Hardy & Phillips, 1999). For example, when people use the concept of health, they create objects such as a healthy or sick person. Note that a healthy or sick person is not just an idea; it has a physical existence. Moreover, very different material consequences follow the construction of a healthy versus a sick person (Foucault, 1973). Finally, "subject positions" refer to the roles or positions that one can take as a participant in discourse. These roles define and delimit how one can meaningfully engage in discursive practices (Hardy & Phillips, 1999; Phillips & Hardy, 1997). For example, an actor who participates in healthcare discourse has already taken one or more of the subject positions such as healthy, sick, doctor, patient, expert, layperson, caregiver, and caretaker. The contributions that a doctor can make to the healthcare discourse are very different from that of a patient (Noone & Stephens, 2008).

As a research orientation, CDA can be particularly fruitful in delineating the discursively constructed distinction between the digital and the physical work environments. By stripping away the objective appearance of this distinction and delving into its underlying performative circuits (Shadnam, 2019), CDA can shed light on the practices of subjectivity in creating the concepts of the physical, remote, and hybrid work (Halford & Leonard, 2006). The challenging aspects of remote and hybrid work such as sustaining the corporate culture (Hirsch, 2021) or maintaining fairness between in-person and remote employees (Mortensen & Haas, 2021) can then be explored and explicated through analysis of the discourses that embody culture, ethics, and power relations.

5.2   Argument

Workplace researchers have understood and appropriated CDA in a wide variety of different ways leading to a confusing literature where "discourse" does not have an agreed-upon meaning or referent, and "analysis" is conducted in eclectic, idiosyncratic ways (Alvesson & Kärreman, 2000). For example, while some workplace researchers take a Foucauldian stance where discourse dominates everything and constitutes objects and subject positions (e.g., Vaara & Tienari, 2002), other researchers give agency a prominent role in the constitution and strategic use of discourse (e.g., Hardy et al., 2000). Lack of critical reflection upon this kind of conceptual heterogeneity has led to a great deal of confusion and inconsistency in the literature. As a result, some scholars called for increased methodological rigor and disciplined use of vocabulary in the studies that employ CDA (Alvesson & Kärreman, 2011; Leitch & Palmer, 2010). Other scholars have criticized these calls by arguing that strict methodological protocols hinder the capacity of CDA to be formulated in contingent ways within engagements with conceptual and empirical problems (Chouliaraki & Fairclough, 2010; Phillips & Oswick, 2012).

In this chapter, I introduce a middle-ground perspective that clearly articulates what CDA is and when and how workplace researchers can employ it, but also highlights the methodological elasticity of CDA in practice. I argue that although "CDA does not have a fixed theoretical and methodological position" (Wodak, 2011, p. 40), avoiding disciplined debate about the range of theoretical and methodological positions in CDA would lead toward further mystification and inaccessibility of CDA for researchers. The critical aims of CDA to have an impact on the pressing social problems of our time and to make a difference cannot be realized unless CDA seizes to be presented as a vague orientation and offers explicit and practical guidelines for dealing with empirical fields. At the same time, I argue against excessive narrowing of CDA and boxing it into a limited set of methodological protocols, which would go counter to the broad scope and agenda of CDA. In the next section, I provide a generic process model that highlights three broad stages of CDA. This model helps workplace researchers by figuring the boundaries of the methodological elasticity of CDA and providing with a skeleton for a research design based on CDA (Denzin, 2009). Finally, I discuss the implications of CDA for workplace researchers and practitioners.

5.3 The generic process model of CDA interwoven with an example

The extant literature of CDA shows that the two fundamental characteristics identified in the first section have shaped a generic process for practicing CDA studies. I present this process in terms of a model for CDA that embarks on a journey (1) from context to text, (2) dives deep into a set of texts, and then (3) comes back to the context. The process is characterized by oscillation between context and text. Researchers begin with what presents itself as a challenging problem, and a rough understanding of its context. This leads them to poke around the set of texts that has fixated that problem, and further analyzing a sample of those texts. The textual analysis then reveals new insights for understanding the context and the levers that can be employed to overcome the selected problem. See Figure 5.1 for an overview of the three stages and the core questions that can guide the analysis in each stage.

For illustration, I use Breeze's study (2012) of the discourses of legitimation in the annual reports of oil corporations following the catastrophe in the Gulf of Mexico in 2010. In this case, the author employed CDA to study how capitalist actors seek to legitimize their existence and practice even when the drastic consequences of their operation lead to an environmental disaster. This example shows how a spatially and materially bounded crisis in a physical workplace environment (an ultra-deepwater offshore drilling rig) has a significant semiotic dimension in the broader social context, and how workplace management practices extend their reach into the management of that semiotic dimension. I illustrate that the steps taken in Breeze's study nicely follow the generic model presented in this chapter.

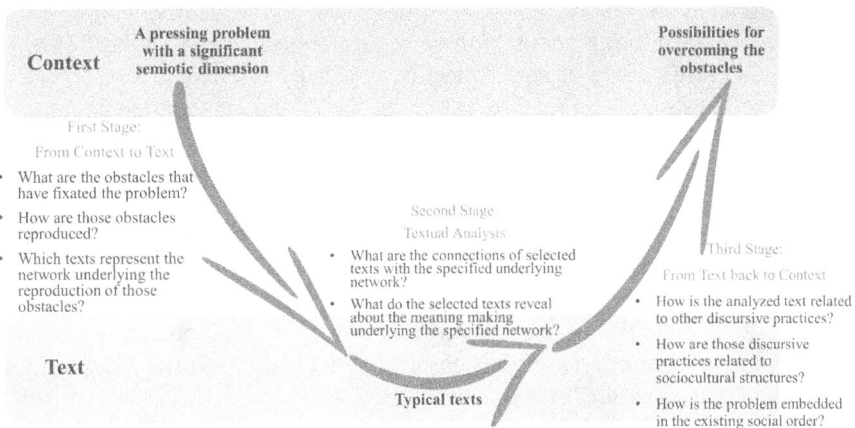

Figure 5.1 The generic process model of CDA

5.3.1 First stage: From context to text

CDA starts from a pressing problem with a significant semiotic dimension in the broader social context (Chouliaraki & Fairclough, 1999). The critical aspect of CDA implies that from square one the researcher is aiming to make a difference in the world rather than contemplating what is already accepted and normalized. The discourse aspect of CDA implies that this problem is closely related to language and semiosis. The research practice of CDA thus begins by focusing on an important social problem, one that does not accept an easy off-the-shelf solution and has a semiotic dimension. Serious social problems of this sort are most often related to power abuse or domination. So CDA starts by aiming at a social problem "with an attitude of opposition and dissent against those who abuse text and talk in order to establish, confirm or legitimate their abuse of power" (van Dijk, 2001, p. 96).

From this point in the context, the analysis starts. The first question in front of the researcher is why the problem is hard to resolve. By examining the immediate scenarios for resolving the problem, the researcher identifies the set of obstacles that have fixated the problem. In other words, these are the obstacles that have made the problem difficult to tackle. The analysis then focuses on these obstacles and investigates the sociopolitical and historical context of the situation. The analysis of the context allows identifying networks of texts, discursive practices, and extra-linguistic social practices that reproduce those obstacles. By explicating these networks and their relation to the problem at hand, the researcher pinpoints and collects a set of texts that represents the problematic character of the specified networks for closer analysis. This set of texts is usually comprised of "typical" texts; i.e., they are representative of a larger body of texts (Wodak, 2001). Here, the term "text" is used broadly to refer to any symbolic expression. So it is inclusive of what is inscribed in spoken, written, or otherwise depicted forms. To collect a set of texts that are typical or representative, CDA researchers do not use any fixed guideline; rather they treat text collection as an ongoing process informed by their theoretical standpoint: "after the first collection exercise it is a matter of carrying out the first analyses, finding indicators for particular concepts, expanding concepts into categories and, on the basis of these results, collecting further data" (Meyer, 2001, p. 24). This mode of data collection is usually referred to as "theoretical sampling" (Glaser & Strauss, 1967; Strauss, 1987) and can also be assisted by computerized operations (Wolfe et al. 1993).

In the case of Breeze's article (2012), for example, the broad social problem that is targeted is the existing political and economic system that is often referred to as "late capitalism" (Mandel, 1975). One major obstacle that has fixated this problem is "the ideological legitimations of late capitalism, as the system itself and agents within that system seek to justify themselves in an increasingly complex scenario" (Breeze, 2012, p. 16). Breeze then narrowed down her attention and focused her analysis on the manifestation of this obstacle in the legitimation of large corporations. She argued that the "study of the messages issued by major corporations is important because it sheds light on the way powerful actors seek to sway public opinion in order to avoid being discredited and to preserve their freedom of action"

(p. 5). Accordingly, Breeze offered a discussion of various networks of texts, discursive practices, and social practices that underpin the ideological legitimations of corporations. In terms of texts, for instance, she noted how the legitimation of an institution entails "asserting the rightness of the institution's actions, preempting criticism from other social players, and marking out the boundaries of the institution's rights and obligations within society" (p. 4). In terms of discursive practices, she described how "non-state entities, such as interest groups, institutions, and corporations, increasingly participate in this type of self-justificatory activity, making statements in the form of press releases, corporate publications, brochures, etc." and also how "these messages reach the public directly, but may also be taken up selectively by the media to operate through mediated channels" (p. 5). Finally, in terms of social practices, she discussed the imposition of fines or punitive damages on corporations as well as the investment decisions of shareholders (p. 6). To better explore the semiosis involved in the legitimations of corporations, she chose to focus on "the messages to shareholders issued by the presidents and Chief Executive Officers (CEOs) of five leading US and European petroleum companies published in the annual reports for the year 2010" (p. 6).

5.3.2 Second stage: Textual analysis

At the end of the first stage, the researcher arrives at one or more sample texts that represent the central role of semiosis in the specified obstacles. The second stage is focused on the linguistic analysis of these sample texts. It starts from delineating the connections of the sample texts with the specified networks of texts, discursive practices, and social practices. More knowledge about the sample texts leads to a better characterization of those texts in the broader picture, which would lead to a better understanding of the role and function of those texts. So although this stage is primarily concerned with textual analysis, it never loses sight of context. CDA always oscillates between text and context (van Dijk, 1993).

The practice of CDA in this second stage is not restricted to any particular method of textual analysis. Rather it can draw on any number of concepts and techniques to study the sample texts in detail and explore their social significance (Wodak, 2011). By the end of this stage, the sample texts are analyzed with respect to their role in meaning-making, thereby producing and reproducing the networks underlying the specified obstacles. The results of this stage can also shed light on a broader array of related semiotic attempts that are aimed at reproducing or disrupting those networks of texts, discursive practices, and social practices.

In the example of Breeze's article (2012), she drew on the work of van Dijk on discursive legitimation (1980, 1993, and 2006) for the textual analysis of the oil corporations' letters to their shareholders. She analyzed these texts and discussed their social significance in the context of the conditions after the Gulf of Mexico disaster. This examination of corporate messages revealed, "evidence of legitimation both of the company itself and the oil industry as a whole" (Breeze, 2012, p. 7). In terms of the legitimation of the oil industry, she highlighted, among other strategies, how "the world's need for energy (framed in glowing terms of growth, progress and transformation) is presented as common sense, and the need for oil appears by direct association as the logical consequence of the former" (p. 9). In terms of the legitimation of individual corporations, she showed several strategies employed by the authorial voice of those texts, including, but not limited to, taking the roles of the financial advisor, expert authority, and environmentally concerned, and, in the case of British Petroleum (BP), positioning the company as a "survivor" of a tragic situation (p. 15).

5.3.3 Third stage: From text back to the context

In this last stage, the researcher completes the analysis of the text by considering the broader context and examining the social problem that initially motivated the study in the new light provided by the textual analysis. This involves the two interrelated processes of interpretation and explanation (Fairclough, 1992; Titscher et al., 2000). "Interpretation" of a text is accomplished through relating the text to the larger ensemble of discursive practices in which it is produced and consumed. This can be accomplished through an analysis of how sets of texts are shaped by prior sets of texts (i.e., the notion of "intertextuality"; see Fairclough, 1992) and how overt and covert aspects of the specified networks of practices are producing, disseminating, and consuming texts. "Explanation" involves relating those discursive practices to other sorts of sociocultural practices. Explanation shows how a text and the larger ensemble of discursive practices that it represents are both shaping and shaped by social structures. All three dimensions of text analysis, interpretation, and explanation are interdependent and "mutually explanatory" (Janks, 1997, p. 26).

The researcher then uses the analysis results at this stage to better understand how the problem at hand is embedded in the existing social order (Fairclough, 2001). This entails enriching our knowledge of the social problem at hand and may even bridge between discursive and extra-discursive aspects of it (Thompson & Harley, 2012). Sometimes, this understanding can also highlight the possibilities for overcoming the obstacles by showing how the opposing or alternative discourses can change the existing networks of texts, discursive practices, and social practices. Borrowing discourses of value from unfamiliar settings is one fruitful approach toward this goal (Keim & Shadnam, 2020).

In the example of Breeze's study (2012), she used her textual analysis to argue that the interweaving of the three discourses of "financial solidity", "scientific expertise", and "environmental concern" in the examined sample of texts is indicative of a kind of "discoursal hybridity" that "is increasingly frequent in corporate genres" (p. 16). She discussed the role of this hybridity in dealing with "unsettling boundaries and disembedding of language practices that characterize contemporary social life" (p. 16). Despite the appearance of not having any stake in politics or social control, she argued that "if we start from the premise that these issues are presented in these particular forms to mesh with the socially generated system of discursive affordances and constraints, then we may perceive that these discourses operate on an ideological level to underpin the workings of late capitalism in an increasingly complex scenario" (p. 17).

5.4 Implications

Here I discuss the implications of the presented model from two perspectives: CDA as research; and CDA as practice:

5.4.1 CDA as research

The implication of the presented model for researchers who employ CDA as a research orientation is twofold: First, I articulated a middle-ground perspective to address the calls for greater rigor in the use of CDA in workplace research (Alvesson & Kärreman, 2011; Leitch & Palmer, 2010; Vaara, 2015) while insisting on the methodological flexibility of this approach (Chouliaraki & Fairclough, 2010; Ferns & Amaeshi, 2019; Phillips & Oswick, 2012). Second, I illustrated that the vast variety of research approaches that fall in the domain of CDA share a generic process inherited from a common intellectual history. This helps in addressing one of the major problems that have been plaguing many of the appropriations of CDA in workplace studies, i.e., the lack of adequate attention to a detailed examination of texts (Aranda et al., 2021; Fairclough, 2005, 2013; Vaara et al., 2010). This can also facilitate the spotting of pattern shifts in collections of texts over time, which can serve as a fruitful foundation for theorizing (Shadnam, 2021).

5.4.2 CDA as practice

The implication for practitioners is that the presented model demystifies CDA and makes it accessible for broad practitioner audiences. It clearly shows that CDA does not necessarily rely on technical and sophisticated textual analysis knowledge, though researchers have employed techniques such as systemic functional linguistics (Askehave, 2010; Iedema, 2003) or corpus linguistics (Mautner, 2005; O'Reilly & Reed, 2011). Instead, it relies on solid analysis of pressing problems with a critical eye. So practitioners can easily adopt and appropriate CDA to change their empirical settings. Moreover, the practitioner's impatience with abstract knowledge and longing for impact in the real world is already a built-in feature of CDA. CDA encourages practitioners to intervene and act, but also equips them with a deep analysis of the problems and the network of factors that have created and maintained those problems. In this vein, a particularly interesting area of workplace studies that can be addressed using CDA is the negotiation and legitimation of different spaces for different products, practices, and purposes (e.g., doing business, philanthropy, activism, etc.). For instance, Johnson et al. (2017) employed CDA to address the contentious debate in France over the inclusion or exclusion of halal meat at a popular burger chain.

5.5 Conclusions

In this chapter, I clarified what CDA is and when and how it can be used by workplace scholars. I highlighted the theoretical specificities of CDA by situating it in the history of its development in linguistics and later workplace studies. I also charted the methodological elasticity of CDA by offering a generic process model for practicing CDA and highlighting a number of questions that guide the researcher in different stages of that process. By doing so, I contribute to previous research and ongoing debates on the use of CDA in workplace research. I particularly articulated a middle-ground approach to address the calls for more rigor in using CDA in workplace research while insisting on the methodological flexibility of this approach. I recalled that CDA has been developed by those linguists whose main common agenda was resolution of important problems in settings ranging from a single social encounter to an organization to the broader society. The related ideological issues, power relations, and the range of methods needed for engaging with those different problems have provided CDA with a great deal of methodological elasticity. However, CDA is of limited value in dealing with problems that are rooted in non-discursive (material) dynamics.

By reviewing the generic process of CDA, my hope is to render it more accessible to workplace researchers while maintaining its internal consistency, rigor, and common principles. The process offered in this chapter provides a clear guiding map for researchers interested in employing CDA in their studies to explicate the role of semiosis in workplaces and hopefully make a difference in it.

5.6   Further Reading

Kaal, B. (2015). How 'real' are time and space in politically motivated worldviews? *Critical Discourse Studies, 12*(3), 330–346. https://doi.org/10.1080/17405904.2015.1013483
Kornberger, M., & Clegg, S. R. (2004). Bringing space back in: Organizing the generative building. *Organization Studies, 25*, 1095–1114. https://doi.org/10.1177/0170840604046312
Lefebvre, H. (1991). *The production of space*. D. Nicholson-Smith (trans). Blackwell.
Richardson, B. (ed.) (2015). *Spatiality and symbolic expression: On the links between place and culture*. Palgrave Macmillan.
Vatansever, F. (2023). From space to spatiality: Critical spatial discourse analysis as a framework for the *geo-graphing* of media texts. *Critical Discourse Studies, 20*(1), 18–35. https://doi.org/10.1080/17405904.2021.1968451

References

Alvesson, M., & Kärreman, D. (2000). Varieties of discourse: On the study of organizations through discourse analysis. *Human Relations, 53*(9), 1125–1149. https://doi.org/10.1177/0018726700539002
Alvesson, M., & Kärreman, D. (2011). Decolonizing discourse: Critical reflections on organizational discourse analysis. *Human Relations, 64*(9), 1121–1146. https://doi.org/10.1177/0018726711408629
Aranda, A. M., Sele, K., Etchanchu, H., Guyt, J. Y., & Vaara, E. (2021). From big data to rich theory: Integrating critical discourse analysis with structural topic modeling. *European Management Review, 18*, 197–214. https://doi.org/10.1111/emre.12452
Askehave, I. (2010). Communicating leadership: A discourse analytical perspective on the job advertisement. *Journal of Business Communication, 47*(3), 313–345. https://doi.org/10.1177/0021943610365310

Berger, P., & Luckmann, T. L. (1966). *The social construction of knowledge: A treatise on the sociology of knowledge*. Doubleday.

Bernstein, B. (1973a). *Class, codes and control* (Vol. 1). Routledge & Kegan Paul.

Bernstein, B. (1973b). *Class, codes and control* (Vol. 2). Routledge & Kegan Paul.

Bernstein, B. (1977). *Class, codes and control* (Vol. 3). Routledge & Kegan Paul.

Breeze, R. (2012). Legitimation in corporate discourse: Oil corporations after deepwater horizon. *Discourse & Society*, *23*(1), 3–18. https://doi.org/10.1177/0957926511431511

Chomsky, N. (1957). *Syntactic structures*. Mouton.

Chouliaraki, L., & Fairclough, N. (1999). *Discourse in late modernity: Rethinking critical discourse analysis*. Edinburgh University Press.

Chouliaraki, L., & Fairclough, N. (2010). Critical discourse analysis in organization studies: Towards an integrationist methodology. *Journal of Management Studies*, *47*(6), 1213–1218. https://doi.org/10.1111/j.1467-6486.2009.00883.x

Denzin, N. K. (2009). *The research act: A theoretical introduction to sociological methods*. Routledge.

Eagleton, T. (1991). The enemy within. *English in Education*, *25*(3), 3–9. doi:10.1111/j.1754-8845.1991.tb01050.x

Fairclough, N. (1992). *Discourse and social change*. Polity Press.

Fairclough, N. (2001). Critical discourse analysis as a method in social scientific research. In R. Wodak & M. Meyer (Eds.), *Methods of critical discourse analysis* (pp. 121–138). Sage.

Fairclough, N. (2005). Discourse analysis in organization studies: The case for critical realism. *Organization Studies*, *26*(6), 915–939. https://doi.org/10.1177/0170840605054610

Fairclough, N., Mulderrig, J., & Wodak, R. (2011). Chapter seven: Critical discourse analysis. In T. A. Van Dijk (Ed.), *Discourse studies: A multidisciplinary introduction* (pp. 357–377). Sage.

Fairclough, N. (2013). *Critical discourse analysis: The critical study of language*. Routledge.

Ferns, G., & Amaeshi, K. (2019). Struggles at the summits: Discourse coalitions, field boundaries, and the shifting role of business in sustainable development. *Business & Society*, *58*(8), 1533–1571. https://doi.org/10.1177/0007650317701884

Foucault, M. (1973). *The birth of the clinic: An archaeology of medical perception*. Vintage.

Glaser, A., & Strauss, B. (1967). *The discovery of grounded theory*. Aldine.

Hardy C., & Phillips N. (1999). No joking matter: Discursive struggle in the Canadian refugee system. *Organization Studies*, *20*, 1–24. https://doi.org/10.1177/0170840699201001

Halford, S., & Leonard, P. (2006). Place, space and time: Contextualizing workplace subjectivities. *Organization Studies*, *27*(5), 657–676. https://doi.org/10.1177/0170840605059453

Hardy, C., Palmer, I., & Phillips, N. (2000). Discourse as a strategic resource. *Human Relations*, *53*(9), 1227–1248. https://doi.org/10.1177/0018726700539006

Hirsch, P. B. (2021). Sustaining corporate culture in a world of hybrid work. *Journal of Business Strategy*, *42*(5), 358–361. https://doi.org/10.1108/JBS-06-2021-0100

Hodgson, G. M. (2001). *How economics forget history: The problem of historical specificity in social science*. Routledge.

Iedema, R. (2003). *Discourses of post-bureaucratic organization*. John Benjamins.

Janks, H. (1997). Critical discourse analysis as a research tool. *Discourse: Studies in the Cultural Politics of Education*, *18*(3), 329–342. https://doi.org/10.1080/0159630970180302

Johnson, G. D., Thomas, K. D., & Grier, S. A. (2017). When the burger becomes halal: A critical discourse analysis of privilege and marketplace inclusion. *Consumption Markets & Culture*, *20*(6), 497–522. https://doi.org/10.1080/10253866.2017.1323741

Jovanović, G. (2011). Toward a social history of qualitative research. *History of the Human Sciences, 24*(2), 1–27. https://doi.org/10.1177/0952695111399334

Keim, C., & Shadnam, M. (2020). Leading in an Amish Paradise: Humanistic leadership in the Old Order Amish. *Cross Cultural & Strategic Management, 27*(3), 473–485. https://doi.org/10.1108/CCSM-01-2020-0024

Labov, W. (1966). *The social stratification of English in New York City*. Center for Applied Linguistics.

Labov, W. (1972). *Sociolinguistic patterns*. University of Pennsylvania Press.

Leitch, S., & Palmer, I. (2010). Analysing texts in context: Current practices and new protocols for critical discourse analysis in organization studies. *Journal of Management Studies, 47*(6), 1194–1212. https://doi.org/10.1111/j.1467-6486.2009.00884.x

Mandel, E. (1975). *Late capitalism*. Verso.

Mautner, G. (2005). The entrepreneurial university. *Critical Discourse Studies, 2*(2), 95–120. https://doi.org/10.1080/17405900500283540

Meyer, M. (2001). Between theory, method, and politics: Positioning of the approaches to CDA. In R. Wodak & M. Meyer (Eds.), *Methods of critical discourse analysis* (pp. 14–31). Sage.

Mortensen, M., & Haas, M. (2021, February 24). *Making the hybrid workplace fair*. *Harvard Business Review*. https://hbr.org/2021/02/making-the-hybrid-workplace-fair

Noone, J. H., & Stephens, C. (2008). Men, masculine identities, and health care utilisation. *Sociology of Health & Illness, 30*(5), 711–725. https://doi.org/10.1111/j.1467-9566.2008.01095.x

O'Reilly, D., & Reed, M. (2011). The grit in the oyster: Professionalism, managerialism and leaderism as discourses of UK public services modernization. *Organization Studies, 32*(8), 1079–1101. https://doi.org/10.1177/0170840611416742

Parker, I. (1992). *Discourse dynamics: Critical analysis for social and psychological analysis*. Taylor & Frances/Routledge.

Phillips, N., & Hardy, C. (1997). Managing multiple identities: Discourse, legitimacy and resources in the UK refugee system. *Organization, 4*(2), 159–185. https://doi.org/10.1177/135050849742002

Phillips, N., & Oswick, C. (2012). Organizational discourse: Domains, debates and directions. *Academy of Management Annals, 6*(1), 435–481. https://doi.org/10.5465/19416520.2012.681558

Shadnam, M. (2019). A postpositivist commentary on self-fulfilling theories. *Academy of Management Review, 44*(3), 684–686. https://doi.org/10.5465/amr.2017.0460

Shadnam, M. (2020). Choosing whom to be: Theorizing the scene of moral reflexivity. *Journal of Business Research, 110*, 12–23. https://doi.org/10.1016/j.jbusres.2019.12.042

Shadnam, M. (2021). New theories and organization research: From the eyes of change. *Journal of Organizational Change Management, 34*(4), 822–837. https://doi.org/10.1108/JOCM-07-2019-0209

Strauss, A. (1987). *Qualitative analysis for social scientists*. Cambridge University Press.

Thompson, P., & Harley, B. (2012). Beneath the radar? A critical realist analysis of 'the knowledge economy' and 'shareholder value' as competing discourses. *Organization Studies, 33*(10), 1363–1381. https://doi.org/10.1177/0170840612457614

Titscher, S., Meyer, M., Wodak, R., & Vetter, E. (2000). *Methods of text and discourse analysis*. Sage.

Turner, B. S. (2000). The history of the changing concepts of health and Illness: Outline of a general model of illness categories. In G. L. Albrecht, R. Fitzpatrick & S. C. Scrimshaw (Eds.), *Handbook of social studies in health and medicine* (pp. 9–23). Sage.

Vaara, E. (2015). Critical discourse analysis as methodology in strategy as practice research. In D. Golsorkhi, L. Rouleau, D. Seidl, & E. Vaara (Eds.), *Cambridge handbook of strategy as practice* (pp. 491–505). Cambridge University Press.

Vaara, E., Sorsa, V., & Pälli, P. (2010). On the force potential of strategy texts: A critical discourse analysis of A strategic plan and its power effects in a city organization. *Organization, 17*(6), 685–702. https://doi.org/10.1177/1350508410367326

Vaara, E., & Tienari, J. (2002). Justification, legitimization and naturalization of mergers and acquisitions: A critical discourse analysis of media texts. *Organization, 9*(2), 275–303. https://doi.org/10.1177/1350508402009002912

van Dijk, T. A. (1980). *Macrostructures: An interdisciplinary study of global structures in discourse, interaction, and cognition.* Lawrence Erlbaum.

van Dijk, T. A. (1984). *Prejudice in discourse: An analysis of ethnic prejudice in cognition and conversation.* Benjamins.

van Dijk, T. A. (1993). Principles of critical discourse analysis. *Discourse & Society, 4*(2), 249–283. https://doi.org/10.1177/0957926593004002006

van Dijk, T. A. (2001). Multidisciplinary CDA: A plea for diversity. In R. Wodak & M. Meyer (Eds.), *Methods of critical discourse analysis* (pp. 95–120). Sage.

van Dijk, T. A. (2006). Discourse, context and cognition. *Discourse Studies, 8*(1), 159–177. https://doi.org/10.1177/1461445606059565

Wodak, R. (2001). What CDA is about – A summary of its history, important concepts and its developments. In R. Wodak & M. Meyer (Eds.), *Methods of critical discourse analysis* (pp. 1–13). Sage.

Wodak, R. (2011). Critical discourse analysis. In K. Hyland & B. Paltridge (Eds.), *The continuum companion to discourse analysis* (pp. 38–53). Continuum.

Wolfe, R. A., Gephart, R. P., & Johnson, T. (1993). Computer facilitated qualitative data analysis: Potential contributions to management research. *Journal of Management, 19*(3), 637–660. https://doi.org/10.1016/0149-2063(93)90008-B

6 Diary studies

Capturing real-time experiences in the workplace

Roman Soucek[1], *Clara Weber*[2], *Jennifer Gunkel*[3] *and Barbara Degenhardt*[4]

[1]*MSH Medical School Hamburg, Germany*

[2]*Zurich University of Applied Sciences, Switzerland & University of Surrey, Great Britain*

[3]*Hochschule Fresenius, Germany*

[4]*University of Zurich, Switzerland*

6.1 Background

The diary research method involves regular, predominantly quantitative self-reports over an extended period with the aim to capture experiences or phenomena whilst near their time of occurrence (e.g., Bolger & Laurenceau, 2013; Iida et al., 2012; Nezlek, 2020; Reis & Gable, 2000). Depending on the research field and era, diary research methods have also been referred to as *intensive repeated measures, intensive longitudinal methods, experience sampling,* or *ecological momentary assessment* (cf. Iida et al., 2012; Nezlek, 2020). In its structured form, diary research methods originated from the social sciences (primarily psychology and anthropology), but humanities (history and literature) were early adopters and used individual diary accounts as historical records to gain insight into eras, events, and "Life as it is lived" (Bolger & Laurenceau, 2013, p. 1; Iida et al., 2012). Diary studies are useful for studying phenomena/processes that rely on the integrity and authenticity of a real-life setting that cannot be studied in a laboratory and when the study focus is on variability in the environment (Ebner-Priemer & Kubiak, 2007; Nezlek, 2020). Diary studies can unveil the fluctuating nature of thoughts, feelings, and behaviours, their antecedents, and their dependence upon situational conditions. As such, this method drastically increases accuracy in formulating and testing hypotheses (e.g., Ohly et al., 2010). As organizational research has shown, there is an astonishing number of organizational and workplace phenomena, often assumed to be stable and hence assessed with cross-sectional methods, that are fluctuant and/or context-dependent (Bissing-Olson et al., 2015; Ohly et al., 2010). These include, for example, work engagement (e.g., Xanthopoulou et al., 2009), work performance (e.g., Binnewies et al., 2009), pro-environmental behaviour (e.g., Bissing-Olson et al., 2015) but also emotional states (e.g., Zohar et al., 2003; cf. Ohly et al., 2010 for a review of diary studies in organizational research). The happy, productive workers' thesis provides an

DOI: 10.1201/9781003289845-6

illustrative example as listed by Ohly et al. (2010) (Fisher, 2003). The thesis is supported by a strong relationship between workers' job performance and job satisfaction when studied longitudinally on an individual level by comparing an individual's job performance at times when they are feeling satisfied to times when they are not feeling satisfied (within-subject approach; Fisher & Noble, 2004). However, the thesis is rather unsupported when studied cross-sectionally on a group level by comparing individuals who feel satisfied to those who do not (between-subject approach; Thoresen et al., 2004). As such, diary studies are useful in uncovering dynamic processes between and within individuals in a specific context, such as in an organization or the workplace. Since diary studies contain daily assessments, they could be classified as longitudinal studies. However, their main scope is not on long-termed developments but on daily fluctuations. Therefore, they comprise repeated and frequent assessments of the same constructs. Ideally, they do not contain any retrospective assessment of certain periods, but the assessment aims at the current experience.

6.2   Argument

6.2.1 *Diary study designs*

Over the past four decades in which diary studies have been used in social sciences, it has been increasingly refined and evolved, especially with the implementation of new technologies (cf. Iida et al., 2012). As such, its application in organizational research has increased over the past few years, especially on research topics concerning occupational health and well-being (cf. Ohly et al., 2010 for an overview). Different types of diary study designs exist, and whilst there are different categorizations (daily diaries, experience sampling, and event sampling, cf. Conner & Lehman, 2012; Ohly et al., 2010), the conservative categorization differentiates three types: time-based (also interval contingent or experience sampling), event-based (also event contingent or event sampling), and device-contingent designs (cf. Iida et al., 2012; Wheeler & Reis, 1991).

Time-based methods collect data according to a time-based log (e.g., every day at noon). *Event-based* methods collect data according to a focal experience of the participant (e.g., social interaction; cf. Rochester Interaction Record; Reis & Wheeler, 1991; Wheeler et al., 1983). *Device-contingent* methods collect data according to changes in participants' physiological condition (e.g., heart rate variability; Baethge et al., 2020) or surroundings (e.g., context-aware experience sampling; Intille et al., 2003) detected by a device. A combination of devices can be used (e.g., e-diaries and wearables; Bernstein & Turban, 2018; Kyriakou et al., 2019). Both the time-based and device-contingent methods can be conducted as *signal-contingent* studies when data collection is prompted by a signal from a device (also referred to as beeper studies or experience sampling; Csikszentmihalyi & Larson, 1987). This can sometimes include random data collection intervals. However, selecting an appropriate method, frequency, and time span for the data collection requires careful consideration and depends on the phenomena and research questions of interest, which will be further elaborated on in the following sections.

6.2.2 Practical considerations concerning study design

Each study design type has its own set of pros and cons: *Time-based* methods use different frequencies (once/a few times a day, once/a few times a week, and so forth) and time spans of data collection (days, weeks, or longer). Frequency and time span "need to be often and long enough to provide a sample of people's lives that is sufficient to provide a basis for making inferences about the topic of the study" (Nezlek, 2020, p. 2). This means, that picking a sensible interval and time span is not only dependent on the research question but also on the concept, process, or phenomena of interest (cf. Iida et al., 2012). For example, some processes change quickly and require shorter intervals for detection than others (e.g., mood, such as workplace satisfaction vs. trait perception, such as personality). This means, that some effects can become undetectable if too much time passes; for example, in coping research, the effect of anxiety is too small to be detected when being assessed a week or longer after coping has occurred (cf. Iida et al., 2012). This could be transferred to workplace research, for example, in a scenario when emotional work outcomes (e.g., emotional exhaustion) are assessed after environmental control perceptions over a socio-environmental stressor (e.g., work privacy invasion). Other effects take longer to manifest and could be missed by using too short intervals (e.g., actual week-to-week changes assessed one week daily). However, too long intervals may lead to the risk of introducing recall bias (Shiffman et al., 2008); recall bias will be less of an issue when data is collected hourly as opposed to once a day. For example, in a study on creativity at work, daily creativity was self-assessed twice a day as creativity is acknowledged to be fleeting (Amabile et al., 2005), whereas overt work behaviours (e.g., proactive behaviour) stay in memory and were assessed once a day (Fritz & Sonnentag, 2009). From a participant's perspective, shorter intervals create a burden for participants; hence the overall investigation period might require shortening. Further, frequencies or intervals chosen must be convenient for participants to ease the burden of regular participation. Concerning time-based methods, the exact timing that is suitable to study processes of numerous phenomena is often unknown; fixed intervals must be decided based on an adequate theory base. If this is not possible, shorter intervals are recommended (Collins, 2006).

Event-based methods require a clear, non-ambiguous definition of the event/ phenomena of interest (e.g., invasion of workplace privacy) and rely heavily on the compliance and judgment of participants. Piloting studies are highly recommended to clarify any questions concerning ambiguous events, e.g., what are the components of work privacy and how does it differ from related concepts such as crowding or territoriality, how do participants know when their work privacy is violated, what are related examples which do not classify as work privacy. With an appropriate definition and reliable registration of the relevant events, the measurement is very close to the immediate experience of these events.

Device-contingent methods come with disadvantages that should be considered when designing a diary study. Reliance on the signalling device requires adequate setup and programming. Further, carrying the device and the frequent signalling disrupt participants' daily routines and may hamper their compliance (Iida et al., 2012). However, regarding the increasing prevalence of smartphones, specialized devices are not necessary in specific cases. For example, the intensity of daily smartphone use could be recorded by the devices themselves, using specified applications installed on the phones. Also, self-monitored health checks, e.g., step counts and pulse counts, have become quite normal and integrated with parts of the population's daily routines since the usage of smartwatches has increased.

Practical considerations and limitations concerning all types of designs include: Requirement of an in-depth briefing of participants; the requirement of a high level of participant commitment to ensure data validity and reliability; participant burden through survey length, frequency, and study length causing poor data and attrition; balancing information yield vs. participant burden; balancing sample size, length, and frequency of assessment vs. resources (money and effort); gauging sample size requirement vs. frequency of assessment/data points (focus on the sample size if interested in between-subject difference; focus on data points if interested in event effects within a person); bias introduced by diary reporting process (e.g., Iida et al., 2012); inability to derive cause-and-effect claims but can detect direction and time-sequence of correlated effects (concerning causal thinking, diary studies are better than cross-sectional studies).

If using questionnaires for data collection, the same questionnaire must be completed multiple times, which challenges the motivation and compliance of the participants (Ohly et al., 2010). Therefore, the questionnaires should be very short and contain a few questions. Regarding these restrictions, scales that consist of more than a few items could be abbreviated by selecting items with the highest item-total correlation (cf. Ohly et al., 2010).

6.2.3 Methodological considerations concerning data analysis

Data analysis of diary studies requires advanced statistical skills. In particular, multilevel analyses are preferred for several reasons. First, missing data at individual points in time do not lead to a large reduction in the sample. If a participant does not complete the questionnaire at a time point, the remaining data of this participant are still considered for the analysis. On the contrary, in analyses of variance for repeated measures, missing data lead to the exclusion of all data of one participant, which quickly reduces the amount of data, especially if many measurement time points are included. This is also an important advantage since diary studies require a high willingness and discipline from the participants (Ohly et al., 2010). For more details: Grund et al. (2019) provided an introduction to handling missing data in multilevel research. Second, since not all participants will complete the questionnaires at the same point in time, time could be mapped with the actual time instead of survey measurement points, which leads to correct interpolations between the time points. For example, a web-based survey could automatically record the time of completion of the questionnaire, and in the statistical analyses, this time goes into "days since the start of the study", which correctly represents time gaps of different lengths between the measurement points. Third, capturing multiple time points allows for modelling curvilinear relationships over time (cf. Ohly et al., 2010 for a detailed discussion of research question examples).

Finally, and of particular importance, multilevel analysis allows the differentiation of between-subject and within-subject effects, for example, by centring on the mean of an individual (within-subject centring; cf. Curran & Bauer, 2011; Singer & Willett, 2003). The differentiation of these effects allows a finer-grained view of influencing variables at different levels. For example, the number of incoming calls may differ daily, and days with many calls may be perceived as more stressful by the same person (within-subject effect). In addition, the workplaces differ regarding the number of incoming calls, as this could be the case for an archivist or a call centre agent (between-subject effect). Furthermore, centring is particularly important when testing for indirect effects from within the multilevel analysis framework (Zhang et al., 2009), which should rely on within-subject effects while statistically controlling for between-subject effects.

Overall, designing a workplace diary study requires consultations with someone experienced in these research methods to help weigh the pros and cons of study design characteristics, flag potential (analytical) problems, and discuss strategies for statistical analyses.

6.3   Examples of application/use

In this section, we present diary studies exemplifying the above considerations concerning study design and data analysis in detail. First, we present a diary study on daily supervisor feedback and its effect on perceived work resources and work engagement. Second, a study on daily interruptions at work addresses questions on different recording methods of events in diary studies. Third, we present two studies on workplace design and well-being that used device-based measurements. Finally, we introduce and discuss an innovative measurement approach, namely the use of pictorial scales.

These examples should provide some inspiration and guidance on applying the methodological approach to one's research. The different methodological approaches to assess daily events or measures such as questionnaires, tally sheets, physiological measurements, or pictorial scales should be noted. Though these are different approaches to measurement, the statistical analyses follow the same pattern, namely distinguishing between persons' mean levels (between-subjects effects) and daily fluctuations around a person's mean level (within-subject effects).

6.3.1 *Assessing daily experiences and consequences: Effects of daily supervisor feedback*

Today's working life is increasingly characterized by virtual, flexible, and self-determined work arrangements. In such circumstances, supervisor feedback can be a powerful job resource and thus a key driver of work engagement. In a diary study, Soucek and Rupprecht (2020) investigated the contribution of day-to-day supervisor feedback concerning job resources and work engagement. The study aimed to investigate the effect of receiving or not receiving supervisor feedback and contrasted two feedback sources, face-to-face feedback versus digitalized feedback, using a computer-mediated feedback system.

Concerning the study design, participants indicated over one month every workday whether they received supervisor feedback and assessed daily job demands as well as how engaged they felt after each day at work. Overall, this diary study consisted of 24 daily repeated questionnaires and was statistically analysed using multilevel models. The daily measurement points were recorded on Level 1 and the participants' reactions on Level 2 (Singer & Willett, 2003). As participants answered the questionnaire only during workdays, this fact results in unequal periods between the measurement times. Therefore, the time variable was assessed in days starting with zero on the first day of the study and later the corresponding day was recorded. This procedure correctly operationalized time, even if the participants answered the daily questionnaires at irregular intervals.

Due to the questionnaire being answered 24 times, the respective constructs are liable to variation between the measurement points (Level 1) and between persons (Level 2). Therefore, individual values for the measured job demands and job resources were centred before being entered as predictors in the multilevel models (within-subject centring; cf. Curran & Bauer, 2011; Singer & Willett, 2003). This procedure resulted in two different predictors. One indicates the average level over the 24 measurement points for every person (between-subjects effect) and the second indicates a person's daily deviations of the person's average value (within-subject effect). This differentiation is particularly important when making statements about the mechanisms behind the systematic change. For example, the between-subjects effect can be influenced by personal characteristics or by attributes of the particular workplace. The within-subject effect contrasts days with higher and lower demands as perceived by the same person. Also, an indirect effect was tested within the framework of multilevel analysis by relating the hypotheses to the within-subject effect at Level 1 while controlling for the between-subjects effect at Level 2 (2-1-1 model; Zhang et al., 2009). In the present study, Soucek and Rupprecht (2020) tested for the indirect effect of supervisor feedback on work engagement using job resources.

6.3.2 Recording of events in diary studies: Daily interruptions at work

The ongoing digitalization and flexibilization of workplaces have expanded and shaped the way of working (Korunka & Kubicek, 2017). Accelerated by the pandemic, communication and collaboration are often media-based, leading to many communication channels and frequent interruptions by emails, instant messages, and video calls. However, assessing interruptions and their consequences is challenging because they distribute throughout a working day and across various media and communication channels (e.g., emails, corporate social networks, and collaboration platforms). Therefore, the assessment of interruptions must rely on recordings by the persons themselves or observations in the workplace. Questionnaires at the end of each working day could be a solution but may be subject to recall bias.

Ebner et al. (2022) investigated the association between daily work interruptions and stress. Thereby, they used two methods to record daily interruptions. In particular, Ebner et al. (2022) prepared a booklet that allowed immediate logging of interruptions with a structured tally sheet that differentiated between various sources of interruption, such as work-related or private interruptions. Participants were instructed to note interruptions as soon as they occurred. In addition to these tally sheets, participants completed a questionnaire on perceived interruptions and indicators of well-being such as psychological detachment.

The study of Ebner et al. (2022) relied on multilevel analyses. The hypotheses were tested on the basis of within-subject effects (i.e., daily interruptions compared to a person's mean level of interruptions) while controlling for between-subject effects (i.e., the mean level of interruptions per person due to each specific work environment). To put it more simple: Days with more interruptions than usual were associated with a lower level of psychological well-being than days with a lower number of interruptions. Furthermore, the results of the diary study by Ebner et al. (2022) indicated that interruptions logged throughout the day were related to the perceived level of interruptions at the end of the day. However, trait-like constructs, such as Fear of Missing Out also influenced the perceived interruptions but not the logged interruptions. This study illustrates the differences between various methods of recording daily events and how the characteristics of the participants influence them.

6.3.3 Device-based measurements in diary studies: Workplace designs and well-being

When conducting workplace research via workshops, observations, interviews, or surveys, workplace users often claim that no week is like the other, no day is like the other, or no hour is like the other. Employees are often not able to report "an average" of time spent in different spaces or distinguish their time spent with different activities and their satisfaction with the work environment to support their activities. To investigate employees' activity, environmental assessments, or mobility profiles, diary methods are easy to use and more precise than letting employees guess about their experiences from a few days ago.

For example, Lindberg et al. (2018) explored the association between workplace design, employee health, and well-being from an occupational health perspective. They measured employees' heart rate variability with a chest-worn sensor, perceived levels of stress via a smartphone app, and physical activity by a triaxial accelerometer sensor on three consecutive workdays. They found that workers in both private offices and traditional, high-partition cubicles were less physically active than workers in open bench seating arrangements to a degree shown to be clinically meaningful in other populations. In addition, higher physical activity levels at the office were clinically meaningful and related to lower physiological stress levels outside the office, indicating careful decisions on workplace layouts and designs to keep sick leaves to a minimum. Also, they found that female workers exhibited significantly lower physical activity levels at the office and higher physiological stress outside the office, pointing to gender-oriented workspace improvements.

In another study, Thayer et al. (2010) examined the effects of physical work environment features on employee stress reactions. Employees' work environment satisfaction and physiological stress working at comparably large workspaces without skylight and no transparent view from the window ('old') were compared with employees' physical and mental reactions working at comparable smaller workspaces but with skylight and transparent windows ('new'). Cardiac activity was measured continuously by using a mobile device. And salvia samples were taken by the employees themselves at certain events during the day to measure cortisol. Finally, the workers reported hourly, prompted by a tone, on the handheld device to what extent, e.g., they had felt stressed, were satisfied with noise/privacy, ventilation, lighting, and views, as well as their consumed units of tobacco, coffee, and alcohol. When the employees answered the first question of each entry of the log, the present time was stored to enable comparison between their questionnaire responses on the smartphone and the cardiac measurements. Using this dairy approach, Thayer et al. (2010) found evidence for greater stress in employees working in a darker workplace with no direct reference to the outside compared to those in lighter workspaces and views with transparent windows. These results can inform management on how to prioritize their investments in a company's work environment, e.g., workspace size vs. lighting and view.

6.3.4 Pictorial scales in diary studies: Daily fluctuation of work intensity

Since diary studies may contain a lot of repeated questionnaires with the same questions, word-based questionnaires might hamper participants' motivation and compliance over time. With diary studies in mind, Soucek and Voss (2022a) developed and validated a questionnaire and pictorial scales on work intensity. The pictorial scales consist of seven pictures representing one of the seven facets of work intensity (Soucek & Voss, 2022a). Each scale comprises five pictures following the idea of a Likert scale. Figure 6.1 shows the pictorial scale of the facet "interruptions at work" which depicts five levels of interruption intensity in five successive pictures. The pictorial scales are available under a Creative Commons License (Soucek & Voss, 2021) and are described by Soucek and Voss (2022b) in more detail concerning their application.

These pictorial scales were used in a diary study that was conducted by a public authority switching from paper-based files to digital files. The study aimed to investigate daily fluctuations in work intensity and their association with job satisfaction. Work intensity was assessed with a series of seven pictures (Soucek & Voss, 2021), and the assessment of job satisfaction relied on a series of circular faces with different mouth curves (Kunin, 1955). The pictorial scales were distributed to employees as printed booklets, including 10 pages for 10 consecutive working days. After the study, the booklets were recollected.

Statistical analyses conducted person-mean centring (cf. Curran & Bauer, 2011; Singer & Willett, 2003) to differentiate between-subjects and within-subject effects. Overall, the pictorial scales evoked sufficient within-subject variance, and

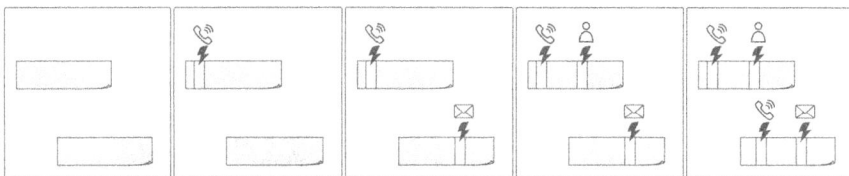

Figure 6.1 Example for a Pictorial Scale: Interruptions (Soucek & Voss, 2021)

results indicated that daily fluctuations of work intensity were associated with daily levels of job satisfaction. Of note, statistically controlling for between-subjects effects utilizing person-mean centring also solved the potential issue of employees' different interpretations of the pictorial scales because testing of the hypotheses relied on the within-subject effects. Overall, the pictorial scales are meeting the demand for a simple, concise instrument that allows for repeated measures, such as in the context of diary studies.

6.4   Implications

6.4.1 Diary method relevance to research

Diary studies can drastically enrich workplace research as it increases rigour in the assessment of phenomena (thoughts, feelings, and behaviours) that are dynamic in their nature, and not stable across contexts and time, which is true for many workplace-relevant phenomena (e.g., employees' satisfaction, performance, or mood). Accuracy can also be gained in intervention studies, as individuals' behaviours (or other reactions) might not have been stable before and might not stay stable after an intervention has occurred. Furthermore, diary studies in workplace research can be combined with other innovative methodological approaches with endless possibilities. For example, investigating the discrepancy between self-rated indoor quality ratings, moderating factors (perception of environmental control), and objective physical measurements (e.g., acoustics). Another area of application could be the assessment of environmental stress (e.g., crowding), moderating factors (e.g., stimulus screening abilities of the individual), and the assessment of physiological stress reactions (e.g., heart rate). Also, the possibility of conducting studies using experience sampling is on the rise (Thai & Page-Gould, 2018).

Overall, diary-style methods allow for more differentiated and accurate investigations (direction and timely order of correlated effects) than prevailing cross-sectional studies. Although only experimental designs reserve the right to derive cause-and-effect relationships, "relative to cross-sectional studies, diary studies are a giant leap forward for causal thinking" (Iida et al., 2012, p. 283). Furthermore, as identified by Ohly et al. (2010), diary methods can advance our understanding in organizational or work research by fostering a process perspective enabling "us to learn more about changing states over time and about how specific states and behaviours translate into other states and behaviours" (p. 85). With diary methods increasing in organizational psychology research and advancing our understanding in this discipline (Ohly et al., 2010), we hope similar effects can be reached in the workplace research discipline.

6.4.2 Diary method relevance to practice

Change management tool. Diary-style methods can aid in getting buy-in from employees and employers for workplace design or change initiatives. First, by reflecting on when, where, and why things are being done, employees can realise that they spend time in different kinds of spaces (not only at their desks). Thus, the 'why' of flexible workplace concepts, including open layouts and desk sharing, can be explained more succinctly. Second, diaries can be used as a self-reflection method and a (self-)coaching tool by which employees of all hierarchy levels can gain new competencies when using spaces. For example, if persons realize doing concentrated work in a crowded space simply out of habit, they can decide to find a better place for this activity. This way of reflecting on user behaviour and workspace experiences can also be done in groups. Thereby, one person's experience can serve as an example of good practice for others.

Tool for continuous organizational learning. Finally, diaries can be used as a tool for the continuous improvement of organizational practices. For example, Becker et al. (2021) found in a diary study conducted in three multinational companies that workspaces in an open office are perceived as less suitable for tasks that demand great concentration. However, for the home office, this did not apply. Companies can use such results to improve their workplace concepts: Where and when should concentrated work be done? Are better or additional spaces required in the office? To evaluate the success of possible changes (like in office design or workplace protocols), diary tools can again generate valuable insights.

6.5 Conclusions

Diary studies can provide a solid empirical foundation for evidence-based decisions in management. Applying a triangulation of different objective and subjective data sources from individuals, the building and indoor environment monitoring, even more, improves the decision basis for future work environment investments (cf. Geng et al., 2019). Especially in piloting expensive building optimization measures and change management, they can accompany organizational interventions and reveal critical mechanisms concerning their effectiveness. Sophisticated statistical analyses differentiate between people and the changes within people over time, including contextual and individual factors impacting these trajectories. However, diary studies are time-intensive and require experienced research competencies with advanced skills in statistical analyses. And the implementation of diary studies requires good preparation to ensure employees' willingness and compliance to participate. In line with Fisher and To (2012), workplace researchers may also plan a single study while addressing several non-overlapping sets of workplace research questions simultaneously.

6.6 Further reading

Full guide to diary style studies: Mehl, M. R., & Conner, T. S. (2012). *Handbook of research methods for studying daily life*. Guilford Press.
A guide on how to conduct and analyse diary style studies: Nezlek, J. B. (2012). Diary methods for social and personality psychology. In J. B. Nezlek (Ed.), *The SAGE library in social and personality psychology methods*. Sage.

A guide on how to conduct and analyse diary style studies with a focus on interval or signal contingent methods: Bolger, N., & Laurenceau, J.-P. (2013). Intensive longitudinal methods: An introduction to diary and experience sampling research. In T. D. Little (Ed.), *Methodology in the Social Sciences*. Guilford Press.

References

Amabile, T. M., Barsade, S. G., Mueller, J. S., & Staw, B. M. (2005). Affect and creativity at work. *Administrative Science Quarterly*, *50*(3), 367–403. https://doi.org/10.2189/asqu.2005.50.3.367

Baethge, A., Vahle-Hinz, T., & Rigotti, T. (2020). Coworker support and its relationship to allostasis during a workday: A diary study on trajectories of heart rate variability during work. *Journal of Applied Psychology*, *105*(5), 506–526. https://doi.org/10.1037/apl0000445

Becker, C., Soucek, R., Gunkel, J., Lütke Lanfer, S., & Göritz, A. S. (2021). Diary study of activity-based flexible offices. *Zeitschrift für Arbeits- und Organisationspsychologie*, *65*(3), 153–164. https://doi.org/10.1026/0932-4089/a000359

Bernstein, E. S., & Turban, S. (2018). The impact of the 'open' workspace on human collaboration. *Philosophical Transactions of the Royal Society B*, *373*(1753), 20170239. https://doi.org/10.1098/rstb.2017.0239

Binnewies, C., Sonnentag, S., & Mojza, E. J. (2009). Daily performance at work: Feeling recovered in the morning as a predictor of day-level job performance. *Journal of Organizational Behavior*, *30*(1), 67–93. https://doi.org/10.1002/job.541

Bissing-Olson, M. J., Fielding, K. S., & Iyer, A. (2015). Diary methods and workplace pro-environmental behaviors. In J. L. Robertson & J. Barling (Eds.), *The psychology of green organizations* (pp. 95–116). Oxford University Press. https://doi.org/10.1093/acprof:oso/9780199997480.003.0005

Bolger, N., & Laurenceau, J.-P. (2013). Intensive longitudinal methods: An introduction to diary and experience sampling research. In T. D. Little (Ed.), *Methodology in the social sciences*. Guilford Press.

Collins, L. M. (2006). Analysis longitudinal data: The integration of theoretical model, temporal design, and statistical model. *Annual Review of Psychology*, *57*(1), 505–528. https://doi.org/10.1146/annurev.psych.57.102904.190146

Conner, T. S., & Lehman, B. J. (2012). Getting started: Launching a study in daily life. In M. R. Mehl & T. S. Conner (Eds.), *Handbook of research methods for studying daily life* (pp. 89–107). The Guilford Press.

Csikszentmihalyi, M., & Larson, R. (1987). Validity and reliability of the experience-sampling method. *The Journal of Nervous and Mental Disease*, *175*(9), 526–536. https://doi.org/10.1097/00005053-198709000-00004

Curran, P. J., & Bauer, D. J. (2011). The disaggregation of within-person and between-person effects in longitudinal models of change. *Annual Review of Psychology*, *62*(1), 583–619. https://doi.org/10.1146/annurev.psych.093008.100356

Ebner, K., Wehrt, W., & Soucek, R. (2022). *"I get interrupted, I get stressed - and I still miss out!" Fear of Missing Out as a moderator of the relationship between daily interruptions at work and stress* [Manuscript submitted for publication].

Ebner-Priemer, U. W., & Kubiak, T. (2007). Psychological and psychophysiological ambulatory monitoring: A review of hardware and software solutions. *European Journal of Psychological Assessment*, *23*(4), 214–226. https://doi.org/10.1027/1015-5759.23.4.214

Fisher, C. D. (2003). Why do lay people believe that satisfaction and performance are correlated? Possible sources of a commonsense theory. *Journal of Organizational Behavior*, *24*(6), 753–777. https://doi.org/10.1002/job.219

Fisher, C. D., & Noble, C. S. (2004). A within-person examination of correlates of performance and emotions while working. *Human Performance*, *17*(2), 145–168. https://doi.org/10.1207/s15327043hup1702_2

Fisher, C. D., & To, M. L. (2012). Using experience sampling methodology in organizational behavior. *Journal of Organizational Behavior*, *33*(7), 865–877. https://doi.org/10.1002/job.1803

Fritz, C., & Sonnentag, S. (2009). Antecedents of day-level proactive behavior: A look at job stressors and positive affect during the workday. *Journal of Management*, *35*(1), 94–111. https://doi.org/10.1177/0149206307308911

Geng, Y., Ji, W., Wang, Z., Lin, B., & Zhu, Y. (2019). A review of operating performance in green buildings: Energy use, indoor environmental quality and occupant satisfaction. *Energy & Buildings*, *183*, 500–514. https://doi.org/10.1016/j.enbuild.2018.11.017

Grund, S., Lüdtke, O., & Robitzsch, A. (2019). Missing data in multilevel research. In S. E. Humphrey & J. M. LeBreton (Eds.), *The handbook of multilevel theory, measurement, and analysis* (pp. 365–386). American Psychological Association.

Iida, M., Shrout, P. E., Laurenceau, J.-P., & Bolger, N. (2012). Using diary methods in psychological research. In H. Cooper, P. M. Camic, D. L. Long, A. T. Panter, D. Rindskopf, & K. J. Sher (Eds.), *APA handbook of research methods in psychology, vol. 1. Foundations, planning, measures, and psychometrics* (pp. 277–305). American Psychological Association.

Intille, S. S., Rondoni, J., Kukla, C., Ancona, I., & Bao, L. (2003, April 5–10). *A context-aware experience sampling tool*. CHI EA '03: CHI '03 Extended Abstracts on Human Factors in Computing System, Ft. Lauderdale, Florida, USA. https://doi.org/10.1145/765891.766101

Korunka, C., & Kubicek, B. (2017). *Job demands in a changing world of work*. Springer.

Kunin, T. (1955). The construction of a new type of attitude measure. *Personnel Psychology*, *8*(1), 65–77. https://doi.org/10.1111/j.1744-6570.1955.tb01189.x

Kyriakou, K., Resch, B., Sagl, G., Petutschnig, A., Werner, C., Niederseer, D., Liedlgruber, M., Wilhelm, F. H., Osborne, T., & Pykett, J. (2019). Detecting moments of stress from measurements of wearable physiological sensors. *Sensors*, *19*(17), 3805. https://doi.org/10.3390/s19173805

Lindberg, C. M., Srinivasan, K., Gilligan, B., Razjouyan, J., Lee, H., Najafi, B., Canada, K. J., Mehl, M. R., Currim, F., Ram, S., Luden, M. M., Heerwagen, J. H., Kampschroer, K., & Sternberg, E. (2018). Effects of office workstation type on physical activity and stress. *Occupational and Environmental Medicine*, *75*(10), 689–695. https://doi.org/10.1136/oemed-2018-105077

Nezlek, J. B. (2020). Diary studies in social and personality psychology: An introduction with some recommendations and suggestions. *Social Psychological Bulletin*, *15*(2), 1–19. https://doi.org/10.32872/spb.2679

Ohly, S., Sonnentag, S., Niessen, C., & Zapf, D. (2010). Diary studies in organizational research. *Journal of Personnel Psychology*, *9*(2), 79–93. https://doi.org/10.1027/1866-5888/a000009

Reis, H. T., & Gable, S. L. (2000). Event-sampling and other methods for studying everyday experience. In H. T. Reis & C. M. Judd (Eds.), *Handbook of research methods in social and personality psychology* (pp. 190–222). Cambridge University Press.

Reis, H. T., & Wheeler, L. (1991). Studying social interaction with the Rochester interaction record. *Advances in Experimental Social Psychology, 24*, 269–318. https://doi.org/10.1016/S0065-2601(08)60332-9

Shiffman, S., Stone, A. A., & Hufford, M. R. (2008). Ecological momentary assessment. *Annual Review of Clinical Psychology, 4*(1), 1–32. https://doi.org/10.1146/annurev.clinpsy.3.022806.091415

Singer, J. D., & Willett, J. B. (2003). *Applied longitudinal data analysis: Modeling change and event occurrence.* Oxford University Press.

Soucek, R., & Rupprecht, A. (2020). Supervisor feedback as a source of work engagement? The contribution of day-to-day feedback to job resources and work engagement. *EWOP in Practice, 14*(1), 70–89.

Soucek, R., & Voss, A. S. (2021). *Pictorial scales on work intensity.* https://doi.org/10.17605/OSF.IO/93KTQ

Soucek, R., & Voss, A. S. (2022a). *Rethinking the assessment of work intensity – Development and validation of a verbal questionnaire and pictorial scales* [Manuscript submitted for publication].

Soucek, R., & Voss, A. S. (2022b). A picture is worth a thousand words: Pictorial scales for the assessment of work intensity. *EWOP in Practice, 16*(1), 45–59.

Thai, S., & Page-Gould, E. (2018). ExperienceSampler: An open-source scaffold for building smartphone apps for experience sampling. *Psychological Methods, 23*(4), 729–739. https://doi.org/10.1037/met0000151

Thayer, J. F., Verkuil, B., Brosschot, J. F., Kampschroer, K., West, A., Sterling, C., Christie, I. C., Abernethy, D., Sollers, J. J., Cizza, G., Marques, A. H., & Sternberg, E. M. (2010). Effects of the physical work environment on physiological measures of stress. *European Journal of Preventive Cardiology, 17*(4), 431–439. https://doi.org/10.1097/HJR.0b013e328336923a

Thoresen, C. J., Bradley, J. C., Bliese, P. D., & Thoresen, J. D. (2004). The big five personality traits and individual job performance growth trajectories in maintenance and transitional job stages. *Journal of Applied Psychology, 89*(5), 835–853. https://doi.org/10.1037/0021-9010.89.5.835

Wheeler, L., & Reis, H. T. (1991). Self-recording of everyday life events: Origins, types, and uses. *Journal of Personality, 59*(3), 339–354. https://doi.org/10.1111/j.1467-6494.1991.tb00252.x

Wheeler, L., Reis, H., & Nezlek, J. B. (1983). Loneliness, social-interaction, and sex-roles. *Journal of Personality and Social Psychology, 45*(4), 943–953. https://doi.org/10.1037/0022-3514.45.4.943

Xanthopoulou, D., Bakker, A. B., Demerouti, E., & Schaufeli, W. B. (2009). Work engagement and financial returns: A diary study on the role of job and personal resources. *Journal of Occupational and Organizational Psychology, 82*(1), 183–200. https://doi.org/10.1348/096317908X285633

Zhang, Z., Zyphur, M. J., & Preacher, K. J. (2009). Testing multilevel mediation using hierarchical linear models: Problems and solutions. *Organizational Research Methods, 12*(4), 695–719. https://doi.org/10.1177/1094428108327450

Zohar, D., Tzischinski, O., & Epstein, R. (2003). Effects of energy availability on immediate and delayed emotional reactions to work events. *Journal of Applied Psychology, 88*(6), 1082–1093. https://doi.org/10.1037/0021-9010.88.6.1082

7 Cluster analysis

Grouping workers by work location choice

Alessandra Migliore and Cristina Rossi-Lamastra

Politecnico di Milano, Italy

7.1 Background

Cluster analysis has been used in quantitative research since the 1970s to group observations into mutually exclusive groups. In so doing, it helps systematize information contained in data (Hair, 2009). Specifically, cluster analysis groups observations through *hierarchical* and *non-hierarchical* algorithms.[1]

Cluster analysis groups observations to maximize *within-cluster homogeneity* and *between-cluster heterogeneity* (Aldenderfer & Blashfield, 1984; Everitt et al., 2011). To assess homogeneity/heterogeneity, all cluster analyses consider measures of dissimilarity/similarity among observations (*e.g.*, the Euclidean distance or correlation matrices). Observations with low or high correlations are assigned to the same cluster, whereas those with high or low correlations are assigned to different clusters. Cluster analysis is well-suited for data exploration. Indeed, the number of clusters is not usually defined *a priori,* and how observations group together emerges *from the bottom up.* Accordingly, the methodology can offer rigorous support to preliminary ideas (*e.g.*, coming from data inspection) on how observations fit together and can be a valuable starting point for hypotheses formulation and testing (Johnson & Wichern, 2002).

Cluster analysis emerged as a meaningful methodology across several disciplines (Hair, 2009). Application domains include (but are not limited to) marketing (*e.g.,* partitioning of consumers into market segments, see Tsiptis & Chorianopoulos, 2009), biological sciences and genomics (*e.g.*, building groups of genes with similar expression patterns, see Eisen et al., 1998), and operation management (*e.g.,* grouping manufacturing strategies across industries, see Frohlich & Dixon, 2001). Cluster analysis has recently been applied to support coding (extensive) corpora of texts to unearth meaningful categories (Namey et al., 2007).

In recent years, workplace research has started to resort to cluster analysis, mainly for grouping (*i.e.,* profiling) workers based on multiple dimensions, including (self-reported) health and comfort (Kim & Bluyssen, 2020), control of indoor climate (Hong et al., 2020), indoor environmental quality (IEQ, Ortiz & Bluyssen, 2022), work motivation (Basińska, 2020), and resignation intention (Wang, 2021). Overall, there are only limited applications of cluster analysis to workplace research. For instance, workplace scholars have rarely used cluster analysis for

DOI: 10.1201/9781003289845-7

profiling workers based on *where* they decide to work and *when* they can adopt multi-local working. The diffusion of *flexible work* (Halford, 2005; Richardson & McKenna, 2013) and of *new ways of working* (Aroles et al., 2019) has put workers' location choices into the limelight, especially in the wake of the COVID-19 pandemic (Mallett et al., 2020). Accordingly, this chapter focuses on two-step cluster analysis and shows its adoption when analyzing workers' location choices.

Two-step cluster analysis uses *hierarchical* and *non-hierarchical* algorithms in tandem (Ketchen & Shook, 1996). Specifically, in the first step, the *hierarchical* clustering procedure developed by Ward (1963) determines the number of clusters and their centroids.[2] In the second step, partitioned or *non-hierarchical* clustering (*e.g.,* k-means clustering) allocates observations to clusters. Scholars (*e.g.,* Frohlich & Dixon, 2001; Ketchen & Shook, 1996) concur that combining the hierarchical and non-hierarchical approaches is more effective than resorting to only one. Accordingly, this chapter illustrates the potential of cluster analysis to advance workplace research on multi-local work. In this case, the unit of analysis is the individual worker and his/her own choices. We also discuss how this technique can help firms and other organizations to profile their workforce based on their work locations' preferences and needs; this profiling, in turn, can support organizations in designing their workplace policies.

7.2   Argument

Even if cluster analysis is a well-established methodology in social sciences, it is not commonly used in research on workplaces and research related to the built environment. Namely, few studies use (two-step) cluster analysis for grouping individuals and their relations with the workspaces. In workplace research, scholars mainly used cluster analysis to define workers' profiles in terms of roles, tasks, and workers' feelings. For instance, Soriano et al. (2020) grouped workers according to two main variables relating to their work type: degree of task complexity and degree of interaction with other people at work. After finding four groups of employees, the study associated each group with its recommended type of space, distinguishing the "fit" (i.e., workers in an adequate office space for their work type) and the "misfit" group (i.e., workers in an inadequate office space for their work type). This association is derived from the predefined assumption that a specific task needs a type of office. Thus, while the study identified *from the bottom-up* groups of employees based on the types of tasks that they perform at work (through cluster analysis), it assigned *a priori* the office environment "adequate" for each group, without considering employees' preferences for doing a certain activity in a certain space. Performing cluster analysis with data on individual choices and preferences could provide useful information on what employees want. This study shows how to apply cluster analysis when dealing with individual choices over work location. Studying workers' location choices is crucial in the contemporary working context. The increasing possibility of choosing work locations – also enabled by Information and Communication Technologies (ICT) – stimulates workers to reflect upon their preferences and needs. At the same time, organizations treasure the outcome of these reflections.

When appointed with autonomy in deciding where to work, workers may choose different locations for their work. Some workers choose to work exclusively at the office; others prefer to work solely from home, while others mix both locations (Halford, 2005; Hislop & Axtell, 2007). The motivation to choose one workspace or the other, or to alternate between them, may depend on different factors, including work tasks and the cost of traveling (*i.e.,* in monetary terms but also in terms of time and effort) (Brown & O'Hara, 2003). Firms and other organizations are interested in workers' preferences and needs regarding work locations because they should decide about their *location flexibility* (*e.g.*, where workers can work, how far from the firm's premises, and when they have to be at the office) to balance their objectives (including those dealing with real estate assets) and those of their workers. For instance, some of the recent workplace flexibility (or inflexibility) policies adopted by firms due to the pandemic have created tensions between workers' quest for flexibility and the organizations' implementation costs.

Organizations address these issues in different ways. Some disregard their workforce's preferences and needs and define their work arrangement policies with a *top-down* approach; a case in point is the recent Tesla CEO's *ultimatum* to his employees to return to the office (Nicholas & Hull, 2021). Other organizations rely on descriptive results obtained from samples of workers and generalize them to the whole workforce, in line with a *one-size-fits-all approach* according to which, supposedly, all workers' preferences and needs coincide with those of the sampled workers. Such systems are dangerous: recent studies (Morning Consult, 2022) report that 55% of remote workers would consider resigning if their firms tried to force them to return to the office. More and more people are quitting their job, a phenomenon popularized as *great resignation*[3] or *great discontent* (Hirsch, 2021), partially motivated by inflexible work arrangements. Although we lack data to understand whether this phenomenon is a short-lived trend – which media amplify – or a long-lasting effect of the COVID-19 pandemic (Ksinan Jiskrova, 2022), it showcases the importance for organizations to make decisions that their workers embrace (Bailey & Rehman, 2022). In such a context, cluster analysis may help decision-makers understand *how many* groups with different preferences and needs emerge from data covering several dimensions (*e.g.*, workers' age, gender, commuting time, and family burden).

Even though other methods exist that can address these issues (*e.g.*, mostly qualitative methods, including focus groups, one-to-one interviews), cluster analysis offers some unique advantages. First, it allows managing large datasets such as the workforce of large firms with many employees and, thus, high heterogeneity in workers' preferences and needs. Second, it enables the replicability of the analysis. Indeed, thanks to widely diffused statistical software packages (*e.g.,* STATA, R, or SPSS), cluster analysis has become an easy-to-run procedure that requires just an adequate knowledge of statistics. Third, partially related to the second point, cluster analysis allows for quick simulations of different scenarios (e.g., what happens when an additional variable is loaded in the analysis). It is worth noting that cluster analysis may return misleading results because of shortcomings in data collection, variable selection, execution of the procedure steps, and tests run on the outcome

(Ketchen & Shook, 1996). In the next section, an example is provided of an application to showcase how two-step cluster analysis works and how to improve its application for future inquiry.

7.3   Example of application/use

This example shows how the authors applied cluster analysis on a large dataset containing information about how Italian academics chose their work locations during the COVID-19 pandemic. The investigation aimed to understand *how many* and *which types* of location choices would emerge for these workers once the strict lockdown in Italy was over (*i.e.,* at the beginning of May 2020, Bontempi, 2021). We expected to find two main groups: those who persisted in working only from home and those who moved between home and the university office. Nonetheless, some surprising evidence is unearthed.

Although some seminal contributions rely on a non-hierarchical algorithm (also known as the k-means or iterative method; see, *e.g.*, Miller & Roth, 1994) where the number of clusters is defined in advance, a two-step clustering procedure is nowadays strongly advised because of its higher validity and reliability (Frohlich & Dixon, 2001; Ketchen & Shook, 1996), especially in workplace research (Soriano et al., 2020). A hierarchical method first determines the number of clusters and cluster centroids; then, it uses them as inputs of the subsequent non-hierarchical algorithm. The following sections explain how to collect and prepare data for the analysis, how two-step cluster analysis groups the observations in the two steps of the procedure, and how to interpret the results.

7.3.1   Data collection

After the first wave of the COVID-19 pandemic (that in Italy ended in May 2020, Bontempi, 2021), a survey was administered via email to the entire population of tenured Italian academics, whose contacts are publicly listed by the Italian Ministry of University and Research (MUR).[4] The target population consisted of 52,630 Italian scholars. Participation in the survey was voluntary and confidential; the survey stayed open from July 24 to September 24, 2020. Among the others, the survey collected data on how often academics worked from multiple work locations (*i.e.,* weekly frequency of access from "never" to "more than five times per week" to the home, the university office, and other third locations of work). The variables included in the cluster analysis were based on these data. *Covid_University* captured the weekly frequency of access to the university for working during the COVID-19 pandemic; *Covid_Home* captured the weekly frequency of access to the home for working during the COVID-19 pandemic; *Covid_Otherspace* captured the weekly frequency of access to other spaces[5] for working during the COVID-19 pandemic.

Noteworthy, there are (in principle) no limitations in the number of variables that one can include in a two-step cluster; however, it is preferable to have a limited number of variables, wisely chosen according to the literature.[6] In the presence of high correlations, one can reduce the number of cluster variables through a principal component analysis (PCA, *e.g.*, Ortiz & Bluyssen, 2022). PCA reduces the correlated variables into fewer independent components, thus, ultimately, solving the problem of multicollinearity[7] (Pacáková & Poláčková, 2013). Ketchen and Shook (1996) criticized PCA because it drops the components with low eigenvalues (a measure of the amount of variance explained by a member). The excluded components may provide unique, important information, and this exclusion may result in a sub-optimal set of clusters. Following Ketchen and Shook (1996), we advise repeating cluster analysis by trying multiple methods for addressing multicollinearity (*e.g.*, PCA or variable standardization[8]) to see how each method may differently affect the results.

7.3.2 Data analysis

We received 11,634 responses, which required cleaning to get rid of incomplete responses. Finally, we obtained 7,865 usable and consistent answers (response rate: 14.94%). The sample included 3,853 women (48.99%) and 4,012 men (51.01%); respondents were on average 51 years old, work in universities located in the North (48.29%), Centre (25.86%), South (25.85%) of Italy, and belong to many scientific fields. Once we selected the variables, we adopted the two-step cluster analysis.

As a first step, we used the hierarchical cluster procedure developed by Ward (1963) to determine the number of clusters and their centroids. We adopted Ward's partitioning and squared Euclidean distance because of its robustness and solidity in maximizing within-cluster homogeneity and between-cluster heterogeneity (Aldenderfer & Blashfield, 1984; Basińska, 2020; Everitt et al., 2011; Frohlich & Dixon, 2001). We referred to the Duda–Hart stopping rule and the Calinski–Harabasz pseudo-F yields equivalent clustering to determine clusters' number and centroids.[9] We also visually inspected the dendrogram[10] to confirm the number of resulting clusters during this step. Namely, starting from the top-down diagram, we detected the number of branches, while starting from the bottom-up diagram, we looked at the points of joining of the branches. Both stopping rules and dendrogram inspection suggested the existence of four clusters in our data.

As a second step, we assigned sampled observations to the four clusters through the k-medians non-hierarchical clustering method. The k-median method allows using the vectors of medians of the variables as centroids.[11] This gave us more consistent and reasonable results than using the k-means clustering method, which instead uses vectors of means.

Finally, to check whether original variables significantly differ across clusters, we ran the one-way analysis of variance (ANOVA) for pairwise comparison of means with a Scheffe post-hoc test. This test was crucial to understanding if the clusters were reasonably defined and profile distinct groups of workers. In addition, to check whether academics in each cluster changed their habits because of the COVID-19 pandemic, we performed matched pairs t-tests within each cluster for variables capturing time spent at different work locations before (*Before_University; Before_Home; Before_Otherspace*[12]) and during COVID-19 (*Covid_University; Covid_Home; Covid_Otherspace*).

7.4   Results

After the abovementioned tests, we confirmed the emergence of four clusters grouping Italian academics according to their location choices. After discussion between the authors, we labeled the four clusters as (1) *home-centric*, (2) *between home and university*, (3) *multi-located*, and (4) *university-centric* (Table 7.1). The labels point to each cluster's main features (*i.e.*, the main location that academics accessed). Therefore, the cluster *home-centric* (Cluster 1) includes a large group of academics (4,564 observations, 58.03% of the sample) that worked solely from home in the observed period (on average 5.395 times per week). The cluster *between home and university* (Cluster 2) collected those who balanced their research activity between home and university but rarely use other spaces and covered one-fourth of the academics in the sample (1,187; 25.26%). The cluster *multi-located* (Cluster 3) isolated those who did research from other spaces (on average 4.508 times per week) more often than from home (on average 3.947 times per week) and sometimes accessed also the campus (on average 1.614 times per week); this group consists of a few academics (368, 4,68%). Finally, the cluster *university-centric* (Cluster 4) summed up those working mainly on-campus (on average, 4.786 times per week) – a relatively small percentage of academics (946, 12.03%).

Based on the Scheffe post hoc test results, we found significant differences across clusters for each variable. The same superscript label in Table 7.1 indicates that the variable's mean is not significantly different in various clusters; this happens only in one case (*Covid_Home* in Clusters 2 and 3).

For relevance, we compare the mean frequencies of access to university, home, and other spaces before and during the COVID-19 period through matched pair t-test. This test's results show that the means of all the variables measuring the frequency of access to the different work locations are different before and during-COVID, suggesting that all academics changed their work location choices because of the pandemic (see Figure 7.1).

Important information that emerged from the clusters helped interpret the phenomenon of multi-location work, including clusters' dimensions, and workers'

Table 7.1 Cluster analysis results

Variables	Sample mean	Cluster 1 – Home-centric (n=4,564) Mean	Cluster 2 – Between Home and University (n=1,987) Mean	Cluster 3 – Multi-located (n=368) Mean	Cluster 4 – University-centric (n=946) Mean
Covid_University	1.424	0.255[d]	2.474[b]	1.614[c]	4.786[a]
Covid_Home	4.430	5.395[a]	3.846[b]	3.947[b]	1.188[c]
Covid_Otherspaces	0.491	0.184[d]	0.355[c]	4.508[a]	0.699[b]

Note: Based on ANOVA tests, the means of all the variables are significantly different among clusters at 99%. Note that the highest mean of each variable is labelled with "a," the next highest mean with "b" and "c," and the lowest mean with "d." The same superscript label indicates that the variable's mean is not significantly different in the various clusters.

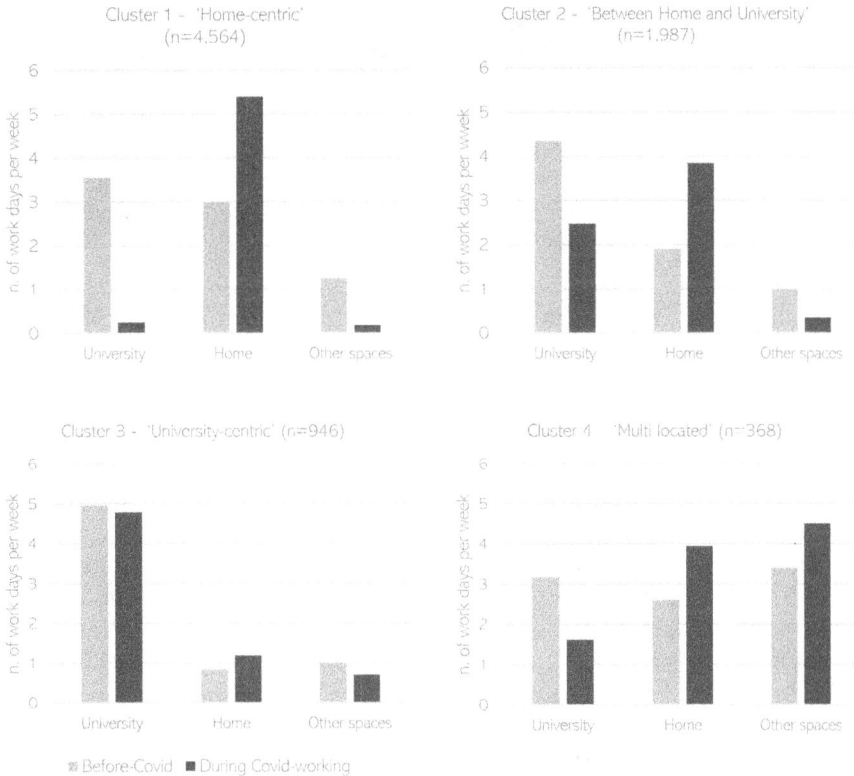

Figure 7.1 Means of frequency of access to university, home, and other spaces before and during COVID-19 in each cluster

distributions among clusters. For instance, academics belonging to the *home-centric cluster* tended to do research from home already about three times per week before the pandemic. On the contrary, those in the university-centric cluster spent limited time at home both during and before the pandemic. This suggests some inertia in work location choices, which may entail home-centric academics being reluctant to return to the university after the pandemic.

Additional analysis – that we cannot describe in this chapter – was necessary to explain the motivations of cluster adherence and advance knowledge of academics' location choices. Therefore, we employed a multivariate analysis to learn more about the demographic characteristics of academics in the four clusters.

7.5   Implications

7.5.1 Method relevance to research

The reported application shows that cluster analysis is highly relevant for workplace research. In the first place, cluster analysis can inform research that aims at profiling categories of users of a workplace to understand their needs, their relations with other users, and their relations with the built environment. Cluster analysis will likely advance workplace research in the future.

First, researchers can easily explore the composition of the clusters (*e.g.*, by statistical tests such as t-test and ANOVA test) to know who belongs to each cluster. For instance, does the cluster include more women than men? More young workers than older ones? More workers engaged in collaborative work than workers carrying on individual tasks? Do employees performing similar work tasks belong to the same cluster?

Second, scholars can understand the temporal evolution of workers' groups by comparing cluster analyses conducted on the same organization in different periods. For instance, they can investigate whether and how clusters of workers tweak after a focal event (*e.g.*, the change in the office location). Clustering workers according to their commuting features (*e.g.*, travel costs, travel time, perceived effort to reach the office, Brown & O'Hara, 2003) sheds light on how many workers likely prefer to work near their main office. Therefore, we believe that the use of cluster analysis contributes to conversations on remote and flexible work.

Third, scholars can compare different organizations by observing how their workers group together according to different measures (*e.g.,* work tasks, work motivations, workplace attachment, etc.). In turn, this might offer interesting insights into heterogeneity in organizational cultures and employees' attachment.

Fourth, by combining cluster analysis with other methodologies (*e.g.,* econometric models), researchers may unearth what influences adherence to the clusters. For instance, one can understand why some workers decide to work from a specific venue and how this choice affects their performances.

Fifth, cluster analysis opens avenues for future methodological advancements in workplace research by overcoming some shortcomings of qualitative methods (*e.g.,* managing large datasets, identifying groups from the bottom up, dealing with subjectivity, and enabling replicability). Along this line of reasoning, scholars can combine cluster analysis with qualitative approaches. This mixed-method approach (Bryman & Bell, 2007) is gaining momentum in social sciences as it allows the depiction of a highly nuanced picture of phenomena.

Finally, cluster analysis may pave new ways of analyzing qualitative data. A telling example consists in coding qualitative text-based data (*e.g.,* data from interviews) into quantitative variables and then applying cluster analysis to these variables (Namey et al., 2007). For instance, one can think of grouping workers according to their experiences in an environment.

7.5.2 *Method relevance to practice*

Cluster analysis results can inform workplace management and corporate real estate strategies. Indeed, amid organizational and environmental constraints, it holds the potential to support organizations in grouping their workforce according to their features, preferences, and needs in a systematic way, rather than relying on anecdotal information.

Profiling workers offers interesting insights to human resources managers (HRMs) who want to promote job autonomy of their workers (Basińska, 2020). It is reasonable to expect that profiling workers according to their motivations improves HRMs' decisions about redesigning the job of some groups (e.g., granting autonomy in task execution).

Likewise, workplace managers' decisions can benefit from cluster analysis because it helps to understand and quantify how different workers' groups may appreciate different work environments, thus ultimately nurturing workers' well-being (Manganelli et al., 2018). For example, it might be wise to locate workers with interdependent tasks on the same floor of an office building or allow those performing independent tasks to work from home. Overall, cluster analysis helps in the design of flexible working policies. This is a timely issue as more and more workplace managers are looking for answers to the following questions: how many days per week can workers work out of the office? Which work locations should they have access to? Should the organization establish satellite offices in some locations (e.g., near where workers reside)?

Finally, firms have been reflecting upon their real estate assets over the last decades. The current adverse macro-economic contingencies (e.g., the increase in electricity prices, the need to reduce CO_2 emissions, the endurance of the COVID-19 pandemic) have urged such reflections and are now calling to action. In such a scenario, grouping employees through cluster analysis is a tool easing the design and implementation of firms' real estate strategies. For example, as the pandemic has made people accustomed to remote working, how much space do firms need for their core activities? How much space could eventually be leased or sold? Where should firms (re)locate their premises to promote workers' well-being and reduce their environmental impacts?

7.6 Conclusions

Nowadays, firms are strongly oriented toward a user-centered approach when designing their workplace and real estate strategies. In such a contest, cluster analysis can be a powerful tool for achieving the goal of understanding in more depth people behaviors. This chapter showcases how cluster analysis – especially the two-step procedure[13] we discussed above – can advance theoretical and empirical research on the workplace in multiple directions. Moreover, it brings an array of exciting practical implications. The robustness and reliability of the cluster analysis results speak in favor of adopting it beyond its mainstream fields of applications (e.g., strategic management, marketing). As any other method, cluster analysis has limitations that open avenues for future research and call relevant stakeholders to take actions.

First, the technique requires a large amount of data. This is undoubtedly a limitation for many workplace researchers and practitioners, given the well-known difficulties in data gathering (e.g., high data collection costs and workers' reluctance to disclose information about their habits). One can overcome data shortage in the short run by resorting to secondary sources (e.g., the European Working Condition

Survey – EWCS[14]). In the long run, we believe that the quest for data will encourage researchers and organizations to develop data collection systems based on cutting-edge technologies (e.g., artificial intelligence, machine learning, and IoT-based solutions). These systems have pros and cons. On the one side, they collect data quickly and automatically, thus reducing the time and effort required to gather data directly from people. On the other side, they pose entangled privacy issues.

Second, adopting cluster analysis requires good statistical skills and tailored software. Accordingly, researchers and practitioners must climb the learning curve by developing new skills and mastering new abilities in computer science. Indeed, implementation mistakes may lead to false evidence and, consequently, to wrong decisions.

7.7 Further Reading

A list of four recommended papers can be found below for those who intend to deepen cluster analysis technique and application. The first paper is one of the seminal works on how to apply cluster analysis and on its main limitations. The other three papers show three applications of cluster analysis on workplace research. While the paper from Basińska groups employees based on work motivation and performance contributing to psychology, the last two papers apply cluster analysis contributing to research in the indoor built environment.

- Ketchen, D., & Shook, C. (1996). The application of cluster analysis in strategic management research: An analysis and critique. *Strategic Management Journal, 17*(6), 441–459.
- Basińska B. A. (2020). Work motivation profiles and work performance in a group of corporate employees: A two-step cluster analysis, *Annals of Psychology, 23*(3), 227–245.
- Kim, D. H., & Bluyssen, P. M. (2020). Clustering of office workers from the OFFICAIR study in The Netherlands based on their self-reported health and comfort. *Building and Environment, 176*, 106860. https://doi.org/10.1016/j.buildenv.2020.106860
- Soriano, A., Kozusznik, M. W., Peiró, J. M., & Mateo, C. (2020). The role of employees' work patterns and office type fit (and misfit) in the relationships between employee well-being and performance. *Environment and Behavior, 52*(2), 111–138. https://doi.org/10.1177/0013916518794260

Notes

1 *Hierarchical algorithms* progress through several steps that build a tree-like structure by either adding observations to (*i.e.*, agglomerative) or deleting them from (*i.e.*, divisive) clusters. The most popular hierarchical algorithms used are agglomerative; they include single linkage, complete linkage, average linkage, centroid method, and Ward's method (Hair, 2009). Selection among the different algorithms depends on the underlying structure of the data (*i.e.*, sample size, distribution of observations, and types of variables included) (Ketchen & Shook, 1996).

Non-hierarchical algorithms (also referred to as k-means or iterative methods) partition a data set into a pre-specified number of clusters (Hair, 2009).

2　The centroid of a cluster is the *center point of the cluster along input variables* (Ketchen & Shook, 1996). For each cluster, it corresponds to the vector of the means of the variables used for cluster analysis.

3　The term Great Resignation was coined by Anthony Koltz, Professor of Management at Mays Business School at Texas, A&M University for describing his prediction of a huge wave of people quitting their jobs in the near future. See the following editorial: Impact of 'The Great Resignation' on organizational knowledge and skills (2022). Business Information Review. https://doi.org/10.1177/02663821221101814

4　These lists include all the Italian scholars tenured in Italian public universities but exclude Ph.D. students, post-doc researchers, and research grant holders — Source of the lists: https://cercauniversita.cineca.it/php5/docenti/cerca.php

5　In line with the extant literature, "other spaces" in our survey include (i) in transit, (ii) at other universities, research centres, labs, or companies, (iii) third spaces as coworking spaces, archives, public libraries, bars, open-air, and parks, and (iv) other environments related to fieldworks or private offices.

6　Of note, we might acknowledge that even if cluster analysis is an explorative methodology, it draws on a post-positivist paradigm that recognizes the relevance of the researcher in shaping scientific knowledge.

7　Usually, the term multicollinearity refers to correlation among the explanatory variables in multivariate regressions (Goldberger, 1991, pp. 245–253). In the case of cluster analysis, it refers to high correlation among cluster variables.

8　Standardization is a procedure of re-scaling variables. Standardization transforms variables so that each has a mean of zero and a standard deviation of one. In the case of cluster analysis, it allows variables to contribute equally to the definition of clusters. However, it may reduce heterogeneity among variables (Ketchen and Shook, 1996).

9　*Duda–Hart* and *Calinski–Harabasz pseudo-F* stopping rules imply the calculation of indices whose values help understand the best number of clusters resulting from the dataset. These stopping rules follow the hierarchical algorithm of the cluster analysis.

　　Calinski–Harabasz pseudo-F stopping rules requires computing the Calinski–Harabasz pseudo-F measure, which looks at the sum of squared distances within the partitions (*i.e.*, data in the same cluster), and compares it to that in the unpartitioned data (*i.e.*, data in different clusters), taking account of the number of clusters and number of cases (Calinski and Harabasz, 1974).

　　The Duda–Hart index is similar to the Calinski–Harabasz pseudo-F. It involves the computation of the Duda–Hart index, which is simply the sum of squares in the two clusters divided by the sum of squares in the combined cluster (Duda et al., 2000).

10　Branching diagram with a tree-like structure showing the relations between observations used in cluster analysis. The dendrogram first groups individual observations and then merges groups until only a unique group is obtained.

11　We used the Stata commands cluster *wardslinkage* and cluster *kmedians* to run the first and second steps of the cluster analysis.

12　*Before_University* captures the weekly frequency of access to the university for working before the COVID-19 pandemic; *Before_Home* captures the weekly frequency of access to the home for working before the COVID19 pandemic; *Before_Otherspace* captures the weekly frequency of access to other spaces for working before the COVID-19 pandemic.

13　As explained in the background section, two-step cluster analysis uses hierarchical and non-hierarchical algorithms in tandem.

14　For further information on the survey see https://www.eurofound.europa.eu/surveys/european-working-conditions-surveys-ewcs

References

Aldenderfer, M. S., & Blashfield, R. K. (1984). *Cluster analysis*. Sage.

Aroles, J., Mitev, N., & de Vaujany, F. (2019). Mapping themes in the study of new work practices. *New Technology, Work and Employment, 34*(3), 285–299.

Bailey, J. R., & Rehman, S. (2022, February 14). How to overcome return to office resistance. *Harvard Business Review*. https://hbr.org/2022/02/how-to-overcome-return-to-office-resistance

Basińska, B. A. (2020). Work motivation profiles and work performance in a group of corporate employees: A two-step cluster analysis. *Annals of Psychology, 23*(3), 227–245.

Bontempi, E. (2021). The Europe second wave of COVID-19 infection and the Italy "strange" situation. *Environmental Research, 193,* 110476. https://doi.org/10.1016/j.envres.2020.110476

Brown, B., & O'Hara, K. (2003). Place as a practical concern of mobile workers. *Environment and Planning A, 35*(9), 1565–1587. https://doi.org/10.1068/a34231

Bryman, A., & Bell, E. (2007). *Business research methods* (2nd ed.). Oxford University Press.

Calinski, T., & Harabasz, J. (1974). A dendrite method for cluster analysis. *Communications in Statistics, 3*(1), 1–27.

Duda, R. O., Hart, P. E., & Stork, D. G. (2000). *Pattern classification* (2nd ed.). Wiley.

Eisen, M. B., Spellman, P. T., Brown, P. O., & Botstein, D. (1998). Cluster analysis and display of genome-wide expression patterns. *Proceedings of the National Academy of Science USA, 95*(25), 14863–14868.

Everitt, B. S., Landau, S., Leese, M., & Stahl, D. (2011). *Cluster analysis* (5th ed.). Wiley.

Frohlich, M., & Dixon, R. (2001). A taxonomy of manufacturing strategies revisited. *Journal of Operations Management, 19*(5), 541–558.https://doi.org/10.1016/S0272-6963(01)00063-8

Goldberger, A. S. (1991). *A course in econometrics*. Harvard University Press.

Hair, J. F. (2009). *Multivariate data analysis* (7th ed.). Prentice Hall.

Halford, S. (2005). Hybrid workspace: Re-spatialisations of work, organisation and management. *New Technology, Work and Employment, 20*(1), 19–33. https://doi.org/10.1111/j.1468-005X.2005.00141.x

Hirsch, P. B. (2021). The great discontent. *Journal of Business Strategy, 42*(6), 439–442. https://doi.org/10.1108/JBS-08-2021-0141

Hislop, D., & Axtell, C. (2007). The neglect of spatial mobility in contemporary studies of work: The case of telework. *New Technology, Work and Employment, 22*(1), 34–51. https://doi.org/10.1111/j.1468-005X.2007.00182.x

Hong, T., Chen, C., Wang, Z., & Xu, X. (2020). Linking human-building interactions in shared offices with personality traits. *Building Environment, 170,* 106602. https://doi.org/10.1016/j.buildenv.2019.106602

Johnson, R. A., & Wichern, D. W. (2002). *Applied multivariate statistical analysis*. Prentice-Hall.

Ketchen, D., & Shook, C. (1996). The application of cluster analysis in strategic management research: An analysis and critique. *Strategic Management Journal, 17*(6), 441–459. https://www.jstor.org/stable/2486927

Kim, D. H., & Bluyssen, P. M. (2020). Clustering of office workers from the OFFICAIR study in he Netherlands based on their self-reported health and comfort. *Building and Environment, 176,* 106860. https://doi.org/10.1016/j.buildenv.2020.106860

Ksinan Jiskrova, G. (2022). Impact of COVID-19 pandemic on the workforce: From psychological distress to the Great Resignation. *Journal of Epidemiol Community Health*, *76*(6), 525–526. https://doi.org/10.1136/jech-2021-218561

Mallett, O., Marks, A., & Skountridaki, L. (2020). Where does work belong anymore? The implications of intensive homebased working. *Gender in Management*, *35*(7/8), 657–665. https://doi.org/10.1108/GM-06-2020-0173

Manganelli, L., Thibault-Landry, A., Forest, J., & Carpentier, J. (2018). Self-determination theory can help you generate performance and well-being in the workplace: A review of the literature. *Advances in Developing Human Resources*, *20*(2), 227–240. https://doi.org/10.1177/1523422318757210

Miller, J. G., & Roth, A. (1994). A taxonomy of manufacturing strategies. *Management Science*, *40*(3), 285–304. http://www.jstor.org/stable/2632800

Morning Consult (2022). New Workers, New Normal. Analyzing the Workforce Transformation Amid COVID-19. *Morning Consult*. https://morningconsult.com/return-to-work/

Namey, E., Guest, G., Thairu, L. N., & Johnson, L. (2007). Data reduction techniques for large qualitative data sets. In G. Guest & K. M. MacQueen (Eds.), *Handbook for team-based qualitative research* (pp. 137–162). Rowman & Littlefield.

Nicholas, K., & Hull, D. (2021). Elon Musk's ultimatum to Tesla execs: Return to the office or get out. *Bloomberg*. https://www.bloomberg.com/news/articles/2022-06-01/musk-s-tesla-ultimatum-return-to-office-or-work-somewhere-else

Ortiz, M. A., & Bluyssen, P. M. (2022). Profiling office workers based on their self-reported preferences of indoor environmental quality and psychosocial comfort at their workplace during COVID-19. *Building and Environment*, *211*, 108742. https://doi.org/10.1016/j.buildenv.2021.108742

Pacáková, Z., & Poláčková, J. (2013). Hierarchical cluster analysis – Various approaches to data preparation. *AGRIS on-Line Papers in Economics and Informatics*, *5*(3), 1–11.

Richardson, J., & McKenna, S. (2013). Reordering spatial and social relations: A case study of professional and managerial flexworkers. *British Journal of Management*, *25*(4), 724–736. https://doi.org/10.1111/1467-8551.12017

Soriano, A., Kozusznik, M. W., Peiró, J. M., & Mateo, C. (2020). The role of employees' work patterns and office type fit (and misfit) in the relationships between employee well-being and performance. *Environment and Behavior*, *52*(2), 111–138. https://doi.org/10.1177/0013916518794260

Tsiptis, K., & Chorianopoulos, A. (2009). *Data mining techniques in CRM: Inside customer segmentation*. Wiley.

Wang, B. (2021). Evaluation method of the excellent employee based on clustering algorithm. In H. N. Dai, X. Liu, D. X. Luo, J. Xiao, & X. Chen (Eds.), *Blockchain and trustworthy systems* (pp 593–600). Springer.

Ward, J. H., Jr. (1963). Hierarchical grouping to optimize an objective function. *Journal of the American Statistical Association*, *58*(301), 236–244. https://doi.org/10.1080/01621459.1963.10500845

8 Stated choice experiments

Identifying workplace preferences and behaviours

Rianne Appel-Meulenbroek and Astrid Kemperman

Eindhoven University of Technology, Netherlands

8.1 Background

Discrete choice (DC) models can be used to describe, explain, and predict the choices people make between two or more discrete alternatives based on their individual preferences. This makes them very helpful in testing alternatives for potential workplace design and/or service interventions. The data that are used as input for these models can be either revealed or stated choices/preferences. Revealed choice/preference data is observed in real-life situations, while stated choice/preference data is based on observations of choices made by individuals under controlled hypothetical situations (Hensher et al., 2015). Thus, revealed choice data is based on past behaviour and is often derived from statistical sources, counts, and observations. Examples in workplace research are observations of actual workspace use in a specific work environment (e.g. Appel-Meulenbroek et al., 2015; Weijs-Perrée et al., 2020). Stated choice data is based on expected future workspace use and -behaviour, if a certain work environment would be provided. The estimated discrete choice models on both types of choice data provide quantitative measures of the relative importance of attributes influencing preferences and choices. In addition, the stated choice approach allows the researcher to include those attributes in the experimental design that are of interest to workplace managers, and to control these attributes and their correlations. Thus, the expected impact of new and even non-existing attributes on choice behaviour and the demand for new products, services, or alternatives can be simulated. This is very interesting when you want to understand and predict preferences for new situations or developments. Specifically, if high investments are required, which changes in a work environment generally need, it is a useful approach.

Among stated choice/preference approaches, a further distinction can be made between the compositional and the decompositional approach. In the compositional approach, respondents are asked to evaluate the attractiveness of each attribute level separately on some rating scale. This approach is actually applied very often in workplace satisfaction research (e.g. Arundell et al., 2018; Maarleveld et al., 2009). By multiplying each attribute's attractiveness and importance scores, one can derive the overall preference for an alternative. This approach has some practical benefits; however, there are also a number of possible problems (Green & Srinivasan, 1990),

DOI: 10.1201/9781003289845-8

such as respondents may not hold all else equal when they provide ratings for the levels of an attribute; social desirability effects may occur when respondents give their rating; respondents in general answer on the basis of their own range of experience over existing options; and importantly, respondents do not have to express trade-offs among attributes. In contrast, the decompositional approach derives the importance weights of attributes from responses to entire choice alternatives based on several attributes. This requires respondents to make a trade-off among attributes, similar to those in real life. Moreover, because an experimental design is used to create hypothetical alternatives, the researcher has control over the attributes and their correlations. Thus, although the respondents are presented with complete alternatives, the use of an orthogonal experimental design allows the researcher to measure each attribute's importance separately in choosing an alternative.

This chapter further focuses on stated choice experiments (SCE) only. In short, respondents are thus asked to choose their most preferred option from a set of exhaustive and mutually exclusive, hypothetical choice alternatives described by their most relevant attributes. Each individual may assess the value of each attribute differently and may also have a different preference for a certain level of an attribute. Individuals integrate their attribute level preferences into an overall utility for each of the alternatives, and it is assumed that they eventually choose the alternative that gives them the highest overall utility (Hensher et al., 2015).

McFadden (1974, 2000) is the founder of discrete choice modelling, and over the years, various aspects of the modelling approach are further advanced by other researchers (e.g. Ben-Akiva & Lerman, 1985; Bhat, 2001; Greene, 2012; Louviere et al., 2000; Train, 2003). The conceptual idea behind individual choice behaviour that underlies discrete choice experiments is derived from various sources, such as Information Integration theory (Anderson, 1970, 1974), Probabilistic choice theory (Luce, 1959), and Random utility theory (Thurstone, 1927). Discrete choice modelling using stated choice experiments has been widely applied in a variety of research fields, such as marketing (e.g. Kaenzig et al., 2013), health (e.g. Johnson et al., 2013), transportation (Greene & Hensher, 2003), and tourism (e.g. Randle et al., 2019). The method has also been applied in several real estate sectors, for example to identify preferences regarding housing (Rouvinen & Matero, 2013; Wang & Li, 2006), retailing (e.g. Balogh et al., 2016), and real estate investment decisions (Del Giudice et al., 2019). It is not very common in the workplace research field, but it has already been used to measure workplace preferences (e.g. Appel-Meulenbroek et al., 2021; Buskermolen et al., 2021; Van den Berg et al., 2020). The next sections argue why stated choice experiments are valuable for workplace research and provide examples of its application for further insight into the method itself.

8.2 Argument

Stated choice experiments have already been used to study how employees judge specific physical workspace attributes in office buildings (Van den Berg et al., 2020), where organisations prefer to locate their office buildings (Balbontin & Hensher,

2021), or how much they would be willing to pay for a healthier office building (Buskermolen et al., 2021). In addition, this method has been used to study other workspaces, such as determinants for willingness to work from home (Ismail et al., 2019) and effects of teleworking on employee perceptions of one another (Mele et al., 2021). It also allows simultaneous research on digital and physical work environment characteristics to identify trade-offs between the two under hybrid working conditions. For example, Appel-Meulenbroek et al. (2022) studied employees' stated preferences for hybrid working, by asking them to choose between several alternative office workspace descriptions and the option to work from home. The model output of this last study provided important insights into which attributes most determined the choice of where to work. It turned out that the expected crowd on the floor and the availability of private spaces for concentration and meetings determined the employees' choices, in this case organisation. That study also estimated a latent class model on the collected data and identified two employee segments with very different preferences: one that intends to re-embrace the office and one that prefers to work from home a lot. Such insights do not only deliver input for theory building on hybrid working, but also help the workplace manager of this case organisation to identify which considerations their employees make when choosing to work in the office versus at home. This helps in adjusting the office to expected hybrid working modes and shows which attributes are most important to focus on if you want your employees to continue to come into the office regularly.

Another advantage of stated choice experiments is that they also allow the introduction of different conditions or contexts under which respondents are asked to make their choice. For example, Appel-Meulenbroek et al. (2022) introduced specific scenarios with amounts of concentrative and/or communication-based activities for a certain day, to choose an appropriate workspace alternative and location. As another example, Tiellemans et al. (2021) introduced an extension of the stated choice technique to estimate residents' preferences that considers group influences and group dynamics in a given group decision context. Specifically, it tested whether people were willing to make concessions to their preferences to reach a group decision. Another third option would be to let people choose between workspace alternatives based on specific desired outcomes such as the ability to concentrate or reduce stress. Such scenarios or contexts can then be included in the data modelling, to get more nuanced insights besides general preferences.

8.3   Example of application/use

Designing a stated choice experiment involves a number of steps (Hensher et al., 2015):

1 selection of influential attributes and their levels;
2 development of the experimental design;
3 constructing the choice task;
4 sampling and data collection procedure; and
5 model estimation.

So far, the method mostly appears to be applied to workplace research by researchers from the Urban Systems and Real Estate group of the Eindhoven University of Technology. They have applied the method to describe and predict behavioural preferences for different types of work environments, such as single tenant offices (Van den Berg et al., 2020), coworking spaces (Appel-Meulenbroek et al., 2021; Weijs-Perrée et al., 2020), and serviced offices (Buskermolen et al., 2021). They also used it on a larger spatial scale to identify location choices of technology-based firms on science parks, including attributes of campus design as a larger scale work environment (Ng et al., 2019). We will explain the study by Van den Berg et al. (2020) in more detail here, as an example of the application of the method and to discuss all the steps necessary in setting up a stated choice experiment (see also Kemperman, 2021 for an overview of steps).

For this example study, we wanted to investigate knowledge workers' preferences for workspaces that support their productivity as much as possible. As the specific type of work activities they need to do might influence workspace preferences, three categories of work activities were introduced as different contexts for these workspace choices: individual concentration work, informal interactions, and formal interactions. First, the most important *attributes and their levels describing the workspaces* were selected from both a knowledge worker and a managerial point of view. If important attributes are missing from the experiment, respondents can make assumptions about the missing ones which can negatively affect the validity of the stated choice experiment. However, the number of attributes included in the experiment needs careful consideration. Including many attributes makes the task too complex for the respondents and complicates the design strategy. On the other hand, including too few might lead to unreliable results. Judging from our experience, about 6–8 attributes seem possible to include in stated choice experiments on workplaces. In addition to the number of attributes, the different levels of each attribute must be defined. One needs to consider that the range should match each respondent's current experience and believability, as the alternatives they must choose from need to be realistic. Moreover, competitive trade-offs should be ensured. For a balanced design, the same number of attribute levels for all attributes is preferred (or for example combinations of 2 and 4 levels). Furthermore, it is important that all combinations of attribute levels are possible and not conflicting with one another. For example, suppose one attribute describes the number of workspaces in a room and is combined with another attribute about noise from roommates. In that case, this could conflict with a possible alternative where there are no roommates, but there is noise from roommates. In such a case, only one of the attributes can be included or the levels should be revised.

To ensure that respondents are not overwhelmed by a large amount of text describing the alternatives, the attribute and the attribute levels are formulated as short and clear as possible. Sometimes it helps to include visuals instead of text for some of the attributes (e.g. Karlsson et al., 2017). One has to be careful though that these visuals only differ on the attribute itself. Therefore, real-life photos of workspaces are generally not suitable, because they can also differ in other aspects. For example, photos from different types of open office sizes might also differ in

Table 8.1 Attributes and their levels

Attribute	Levels
Level of enclosure	Enclosed Semi-enclosed Open
Personal control	Full control over workspace Limited control over workspace No control over workspace
Noise	High noise levels Neutral situation Low noise levels
Ergonomics of the workspace	Specialised ergonomic furniture Ergonomic furniture Non-ergonomic furniture
Temperature	Slightly too cold Neutral situation Slightly too hot
Lighting	Much lighting Pleasant situation Little lighting

lighting levels, the view from the window, or the expression on peoples' faces. So visual attribute levels usually have to be drawn or created specifically for the experiment. Multi-media techniques nowadays provide possibilities such as video and virtual reality to present choice alternatives and situations to the respondents (e.g. Van Dongen & Timmermans, 2019). In our example, six attributes were included, each with 3 levels (see Table 8.1). All attribute levels were presented with qualitative levels (e.g. describing the enclosure of the workplace), but it is also possible to use quantitative levels (e.g. workspace size in m²).

The second step in setting up a stated choice experiment is the development of the *experimental design*. The design should maximise the identification possibilities of the utility function and choice model and the precision of the parameter estimation. In general, the so-called orthogonal designs are still used most, specifically in more applied research. In the last decade, the so-called efficient designs, such as D-efficient designs, D-optimal designs, and Bayesian efficient designs, have been proposed and used (Rose & Bliemer, 2009). These designs are more efficient in generating alternatives and choice tasks that maximise the collected information in the data, yielding more reliable parameter estimates while requiring a lower number of observations than the traditional designs (Van Cranenburgh et al., 2017). However, for these designs, prior parameter estimates are needed as input. Several programs can be used to generate designs such as SAS and Ngene. In this example with 6 attributes that each have 3 levels, a full factorial design would consist of ($3^6 =$) 729 alternative combinations. Therefore, an orthogonal fractional factorial design was used to reduce the number of alternatives to a more reasonable number of 18 alternatives to choose from (see Table 8.2).

Table 8.2 Alternatives using a 3^6 fractional factorial design

Alternatives	Level of enclosure	Personal control	Noise level	Ergonomics	Temperature	Lighting
1	Enclosed	Full control	High	Special ergonomic furniture	Slightly too hot	Little lighting
2	Enclosed	Limited control	Neutral	Non-ergonomic furniture	Neutral situation	Pleasant situation
3	Enclosed	No control	Low	Ergonomic furniture	Slightly too hot	Much lighting
4	Semi-enclosed	Full control	Neutral	Ergonomic furniture	Neutral situation	Much lighting
5	Semi-enclosed	Limited control	Low	Special ergonomic furniture	Slightly too hot	Little lighting
6	Semi-enclosed	No control	High	Non-ergonomic furniture	Slightly too cold	Pleasant situation
7	Open	Full control	Low	Non-ergonomic furniture	Neutral situation	Little lighting
8	Open	Limited control	High	Ergonomic furniture	Slightly too hot	Pleasant situation
9	Open	No control	Neutral	Special ergonomic furniture	Slightly too cold	Much lighting
10	Enclosed	Full control	Low	Ergonomic furniture	Slightly too cold	Pleasant situation
11	Enclosed	Limited control	High	Special ergonomic furniture	Neutral situation	Much lighting
12	Enclosed	No control	Neutral	Non-ergonomic furniture	Slightly too hot	Little lighting
13	Semi-enclosed	Full control	High	Non-ergonomic furniture	Slightly too hot	Much lighting
14	Semi-enclosed	Limited control	Neutral	Ergonomic furniture	Slightly too cold	Little lighting
15	Semi-enclosed	No control	Low	Special ergonomic furniture	Neutral situation	Pleasant situation
16	Open	Full control	Neutral	Special ergonomic furniture	Slightly too hot	Pleasant situation
17	Open	Limited control	Low	Non-ergonomic furniture	Slightly too cold	Much lighting
18	Open	No control	High	Ergonomic furniture	Neutral situation	Little lighting

At which workspace would you work as productive as possible for each of the three work modes indicated below?

Characteristics	Workspace 1	Workspace 2	Neither
Level of enclosure	*Workspace in open environment*	*Workspace in open environment*	
Level of personal control	*No control*	*Full control*	
Noise level	*Neutral situation*	*Low noise levels*	
Furniture	*Special ergonomic furniture*	*Non-ergonomic furniture*	
Temperature	*Slightly too cold*	*Neutral situation*	
Lighting	*Much lighting*	*Little lighting*	

Please provide your choice for:			
Individual concentrated work	o	o	o
Informal interactions	o	o	o
Formal interactions	o	o	o

Figure 8.1 Example of a choice set that was included in the questionnaire

The next step involves dividing the hypothetical workspaces over *choice sets*. The most general approach is to randomly place the alternatives in choice sets. In this example, each respondent evaluated nine choice sets with each two alternatives and a 'no choice option' (see Figure 8.1). Note that in this study the contexts varied by type of activity, and therefore the respondents were asked to select the workspace that supported their productivity best for the three activity types separately.

When the survey is ready, the *sampling and data collection procedure* needs to be determined. This is not different from regular survey approaches and therefore not discussed in detail here. In the example, 14 companies within the authors' networks were found willing to spread the link to the online questionnaire amongst their employees. In total, 251 employees completed the full survey. Although the 14 office-based companies were from many different sectors, they all occupied activity-based offices as the spatial concept that we wanted to study.

After the data have been gathered, one can start the *Model estimation* (see Hensher et al., 2015 for detailed information about modelling approaches). For the model estimation, the dependent variable, the choices made by the respondents, and the independent variables, the attributes, need to be coded. There are two main coding schemes: dummy coding and effect coding. When dummy coding is used, all the attribute levels except one are coded as 1 on their corresponding indicator and 0 on all others. One of the attribute levels is coded as 0 on all indicators. When effect coding is used, attribute levels are coded as 1 on their corresponding indicator, except for one of the attribute levels which is coded as –1 on all indicators. Then, the sum of the effects is equal to zero for each attribute. The overall model fit is the same regardless of the coding scheme used.

The most applied choice model to date is the multinomial logit (MNL) model. However, an important limitation of this model is the Independence from Irrelevant Alternatives (IIA) property, implying that the systematic component of the utility function is a function of only the attributes of the alternative and is independent of the existence and the attributes of all other alternatives in the choice set. However, when it is expected that the choice probabilities of alternatives may be affected by the presence and/or characteristics of other alternatives in the choice set, this may not be desirable. Moreover, the MNL model estimates mean preferences over the complete group of respondents and does not consider heterogeneity in preferences. Over the years, more flexible modelling approaches have been developed such as Latent Class (LC) Models that can be used to find clusters of respondents with similar preferences or Mixed logit (ML) models that can capture heterogeneity in attribute parameters, thus taste differences in preferences between respondents.

In this example, first, a MNL model for each of the work activities was estimated, followed by LC model estimations. The significance level of each attribute level parameter in the output of the MNL shows whether this attribute level significantly influenced the choice of a workspace. Also, for each activity category there is the constant, indicating, in this case, the preference of choosing one of the hypothetical workspaces in the choice set over the 'no choice option'.

The three most important characteristics when choosing a workspace to support the different knowledge worker activities were noise, level of enclosure, and lighting. The main difference between the activity categories was the relative impact of each attribute on workspace choice. In order to make such differences clear, often the relative impacts for all attributes (in this case for the three work activities separately) are presented graphically (see Figure 8.2). Relative importance is calculated by taking the range between the attribute level parameters per attribute, then sum up all the ranges of the attributes, and subsequently dividing the attribute range by the overall sum of ranges. For full insights into the exact model output, parameters, and segments, please check Van den Berg et al. (2020).

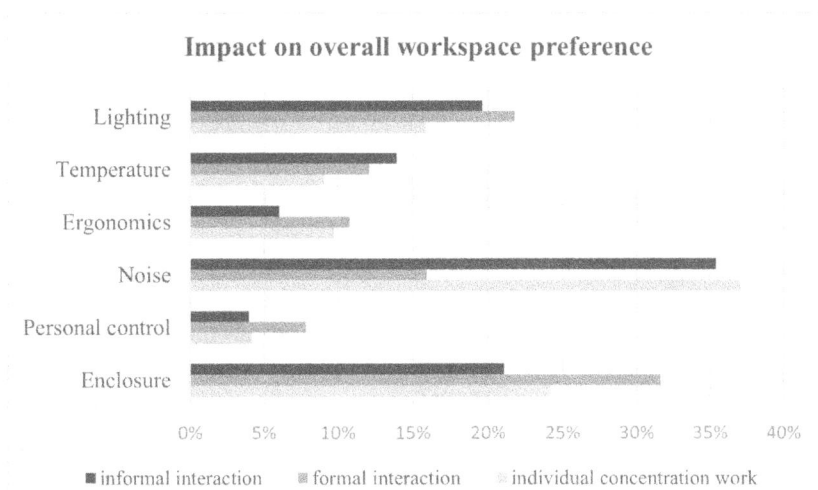

Figure 8.2 Graphical representation of attributes' relative impact on choices

The LC model statistics pointed to a two-class model for informal interactions, while for the work mode formal interactions a three-class model performed better. Generally, the classes that are generated this way are then profiled based on the statistical differences between them and a number of characteristics of the respondents. These statistical differences can be identified with common bivariate statistical approaches such as Chi-square, t-tests, etc., depending on the number of classes and the specific descriptive variables tested (e.g. age, gender, education, job type, etc.). In this case, for example for informal interaction, respondents in class 1 were more likely to choose one of the offered hypothetical workspaces in the experiment and could be called 'non-critics' versus the 'grumblers' in segment 2 that more often choose the 'neither' option. But as age and education level also significantly differed between the classes, the classes were labelled 'young academics' and 'aged critics'.

Due to the low response rates within the 14 different organisations, they would need to repeat the experiment individually with more of their employees, to get valid specific insights for their own context. But the sample was large enough to provide insights for academic theory building. First of all, that noise levels, level of enclosure, and lighting were perceived by these employees as the most important workspace attributes determining their expectations to be productive. In addition, they prefer separate quality levels for different activity scenarios, so ideally the office should provide different areas for different activities. Also, it appears that older employees are more critical in evaluating workspace options than younger ones. It might be that age increases the likelihood of experience with more different types of workspaces, which could make people more aware of what they like and dislike and, thus, more critical.

8.4   Implications

As shown in this chapter, the stated choice experiment approach is suitable and attractive for workplace research. Even though it has only been applied in a very limited number of workplace studies, those studies have shown that the method is usable for many different workplace topics and provides valuable additional possibilities to existing studies and methods.

8.4.1 Method relevance to research

Workplace satisfaction and preferences are the topic of many studies on the effectiveness of work environments for employees and office organisations (Appel-Meulenbroek et al., 2018). However, so far, they have not led to a standard and validated measurement scale on workplace satisfaction. Recent studies (e.g. Kwon & Remøy, 2019; Rolfö, 2018) do not apply measurements scales developed in older works (e.g. Gorgievski et al., 2010; Wagner et al., 2007) but again design their own way of measuring satisfaction. Stated choice experiments can identify preferences for certain design solutions in a very rigorous and structured way. In addition, they can also identify potential behaviours in certain work environments under specific contexts or conditions. Appel-Meulenbroek et al. (2018)'s systematic literature review showed that the ability to perform certain types of activities at certain work-space types is still a major research gap. This method could shed more light on this gap, similar to the examples discussed in this chapter. The trend towards hybrid working has become even more important than before.

Another major benefit is that the hypothetical workplace alternatives shown to respondents can contain any combination of workplace location, design, or use attributes, which is impossible when studying preferences for existing workspaces in real life. So, stated choice experiments add to existing methods the possibility to study workplace design alternatives that have not been built anywhere yet. This way, ideas for future innovations can be evaluated on potential embracement or rejection by future users.

Last, the method is suitable for studying workplace users and workplace managers or line managers. It can be used, for example, to measure the willingness of these decision-makers to implement and/or pay for certain workplace solutions (if the price is included as an attribute).

8.4.2 *Method relevance to practice*

Workplace managers in practice could benefit a lot from future scientific insights as described in section 4.1. When academics get a better grip on general preferences and workspace mechanisms to support people's activities, offices can be improved worldwide to better support employees in their work. But besides benefitting from the scientific progress that stated choice experiments can bring to academic workplace research, the method can also be applied by themselves if they have sufficient statistical skills. It could be a helpful approach for workplace managers to identify preferences or behaviours specifically within their own client organisation, as input for determining future workplace interventions and/or -policies. For example, they could use the method in a participatory design project to identify the preferred design from alternatives that an architect and/or interior designer has come up with for a new-built or renovation project at hand. Instead of asking a small pilot group to look at the drawings, with SCE they can ask all their employees to provide their preferences. If they would add a VR environment to the SCE, employees could experience the intended future design(s) without having to build pilot projects. Another example could be preparing for the organisation to move to another location, where the available options could be tested for the potential acceptance by the employees this way. Basically, any topic that demands an opinion on alternative solutions, options, or strategies can be studied to find out what employees think of it.

8.5 Conclusions

Stated choice experiments are a valuable contribution to workplace research methods because they can identify employee preferences for design alternatives, their behaviours in different types of workspaces, and the willingness to pay and/or accept certain workplace alternatives of both workplace managers and users. Its rigorous and experimental design can add more structure and new insights to existing workplace satisfaction and -use studies. There are several research gaps that this method could be applied to.

This chapter has shown insights in the five main steps to set up a stated choice experiment on any of these topics. Especially step 1 is vital in determining the success of these experiments and the validity of their results (Kløjgaard et al., 2012). Extensive literature studies should be performed to identify which attributes to include and how to divide their design options over different levels. Perhaps the most significant limitation of a stated choice experiment is that only a limited number of attributes can be included. It is not always easy to choose which to include and which not to. When important attributes are not included, this is not visible in the results of a stated choice experiment. Another limitation is that the design is not longitudinal, implying that the data is acquired at one moment in time and that therefore the spirit of time is not included. Also, it is not possible to determine how the preferences of a specific individual might change overtime. And like with regular surveys, the mood of respondents can affect their answers. Last, it is not sure if people will make the same choices that they provided during the stated choice experiment in real-life situations (Hensher et al., 2015).

Nonetheless, the additions and benefits that this method is able to bring to the workplace research field make it qualified for further application in future studies. Such studies could identify, for example, how to balance using a physical and a virtual/home workspace for hybrid workers. Regarding corporate real estate and workplace management in general, the logit models can identify the most important attributes to steer on, the latent class models whether certain segments of workers can be distinguished, and the willingness to pay measures whether certain interventions in the work environment are feasible and desirable in the eyes of decision-makers.

8.6   Further Reading

There are some excellent handbooks written by Hensher et al. (2015) and Louviere et al. (2000) describing all the steps involved in setting up a stated experiment in detail and we would like to refer the readers who are interested in this approach to these books. Specifically, these books can be used by beginners in discrete choice modelling, but also provide insightful knowledge for more experienced researchers. Some basic (or preferably advanced) statistical knowledge might help in understanding and applying all the steps involved in the approach:

Hensher, D. A., Rose, J. M., & Greene, W. H. (2015). *Applied choice analysis* (2nd ed.). Cambridge University Press.
Louviere, J., Henscher, D., & Swait, J. (2000). *Stated choice methods: Analysis and applications*. Cambridge University Press.

In addition to these handbooks, Kløjgaard et al. (2012) describe very extensively how you can select your attributes their quality levels in the best way:

Kløjgaard, M., Bech, M., & Søgaard, R. (2012). Designing a stated choice experiment: The value of a qualitative process. *Journal of Choice Modelling, 5*(2), 1–18.

For other elaborated examples of the application of the method in the workplace field than the one that is discussed in section 3, you can have a look at:

Appel-Meulenbroek, R., Kemperman, A. Van de Water, A., Weijs-Perrée, M., & Verhaegh, J. (2022). How to attract employees back to the office? A stated choice study on hybrid working preferences. *Journal of Environmental Psychology, 81*, 101784.

Buskermolen, W., Appel-Meulenbroek, R., Arentze, T., & Kemperman, A. (2021, June 5–8). Willingness to pay for healthy office workplace aspects. In *Proceedings of the 27th annual ERES conference* Online event. https://library.eres.org/eres2021/paperupload/P_20210520091936_9579.pdf

References

Anderson, N. H. (1970). Functional measurement and psychophysical judgement. *Psychological Review, 77*(3), 153–170.

Anderson, N. H. (1974). Information integration theory: A brief survey. In D. H. Kraantz, R. C. Atkinson, R. D. Luce & P. Suppes (Eds.), *Contemporary developments in mathematical psychology* (Vol. 2, pp. 236–301). Freeman.

Appel-Meulenbroek, R., Clippard, M., & Pfnür, A. (2018). The effectiveness of physical office environments for employee outcomes: An interdisciplinary perspective of research efforts. *Journal of Corporate Real Estate, 20*(1), 56–80. https://doi.org/10.1108/JCRE-04-2017-0012

Appel-Meulenbroek, R., Kemperman, A., Kleijn, M., & Hendriks, E. (2015). To use or not to use: Which type of property should you choose? Predicting the use of activity-based offices. *Journal of Property Investment & Finance, 33*(4), 320–336. https://doi.org/10.1108/JPIF-09-2014-0059

Appel-Meulenbroek, R., Kemperman, A., Van de Water, A., Weijs-Perrée, M., & Verhaegh, J. (2022). How to attract employees back to the office? A stated choice study on hybrid working preferences. *Journal of Environmental Psychology, 81*, 101784. https://doi.org/10.1016/j.jenvp.2022.101784

Appel-Meulenbroek, R., Weijs-Perrée, M., Orel, M., Gauger, F., & Pfnür, A. (2021). User preferences for coworking spaces: A comparison between the Netherlands, Germany and the Czech Republic. *Review of Managerial Science, 15*(7), 2025–2048. https://doi.org/10.1007/s11846-020-00414-z

Arundell, L., Sudholz, B., Teychenne, M., Salmon, J., Hayward, B., Healy, G. N., & Timperio, A. (2018). The impact of activity based working (ABW) on workplace activity, eating behaviours, productivity, and satisfaction. *International Journal of Environmental Research and Public Health, 15*(5), 1005. https://doi.org/10.3390/ijerph15051005

Balbontin, C., & Hensher, D. A. (2021). Understanding business location decision making for transport planning: An investigation of the role of process rules in identifying influences on firm location. *Journal of Transport Geography, 91*, 102955. https://doi.org/10.1016/j.jtrangeo.2021.102955

Balogh, P., Békési, D., Gorton, M., Popp, J., & Lengyel, P. (2016). Consumer willingness to pay for traditional food products. *Food Policy, 61*, 176–184. https://doi.org/10.1016/j.foodpol.2016.03.005

Ben-Akiva, M., & Lerman, S. R. (1985). *Discrete choice analysis: Theory and applications to travel demand.* MIT Press.

Bhat, C. R. (2001). Quasi-random maximum simulated likelihood estimation of the mixed multinomial logit model. *Transportation Research Part B, 35*(7), 677–693. https://doi. org/10.1016/S0191-2615(00)00014-X

Buskermolen, W., Appel-Meulenbroek, R., Arentze, T., & Kemperman, A. (2021, June 5–8). *Willingness to pay for healthy office workplace aspects* [Paper presentation]. European Real Estate Society Conference, Online event. https://library.eres.org/eres2021/paperupload/ P_20210520091936_9579.pdf

Del Giudice, V., De Paola, P., Francesca, T., Nijkamp, P. J., & Shapira, A. (2019). Real estate investment choices and decision support systems. *Sustainability, 11*(11), 3110. https://doi. org/10.3390/su11113110

Gorgievski, M. J., Van der Voordt, T. J., Van Herpen, S. G., & Van Akkeren, S. (2010). After the fire: New ways of working in an academic setting. *Facilities, 28*(3/4), 206–224. https://doi.org/10.1108/02632771011023159

Green, P., & Srinivasan, V. (1990). Conjoint analysis in marketing: New developments with implications for research and practice. *Journal of Marketing, 54*(4), 3–19. https://doi. org/10.2307/1251756

Greene, W. H. (2012). *Econometric analysis* (7th ed.). Prentice Hall.

Greene, W. H., & Hensher, D. A. (2003). A latent class model for discrete choice analysis: Contrasts with mixed logit. *Transportation Research Part B: Methodological, 37*(8), 681–698. https://doi.org/10.1016/S0191-2615(02)00046-2

Hensher, D., Rose, J., & Greene, W. (2015). Experimental design and choice experiments. In *Applied choice analysis* (pp. 189–319). Cambridge University Press. https://doi. org/10.1017/CBO9781316136232.008

Ismail, F. D., Kadar Hamsa, A. A., & Mohamed, M. Z. (2019). Modelling the effects of factors on the stated preference towards telecommuting in IIUM campus. Gombak. *International Journal of Urban Sciences, 23*(1), 122–147. https://doi.org/10.1080/ 12265934.2018.1446352

Johnson, F. R., Lancsar, E., Marshall, D., Kilambi, V., Mühlbacher, A., Regier, D. A., Bresnahan, B. W., Kanninen, B., & Bridges, J. F. P. (2013). Constructing experimental designs for discrete-choice experiments: Report of the ISPOR conjoint analysis experimental design good research practices task force. *Value in Health, 16*(1), 3–13. https://doi. org/10.1016/j.jval.2012.08.2223

Kaenzig, J., Heinzle, S. L., & Wüstenhagen, R. (2013). Whatever the customer wants, the customer gets? Exploring the gap between consumer preferences and default electricity products in Germany. *Energy Policy, 53*, 311–322. https://doi.org/10.1016/j. enpol.2012.10.061

Karlsson, L., Kemperman, A., & Dolnicar, S. (2017). May I sleep in your bed? Getting permission to book. *Annals of Tourism Research, 62*, 1–12. https://doi.org/10.1016/j. annals.2016.10.002

Kemperman, A. (2021). A review of research into discrete choice experiments in tourism - Launching the annals of tourism research curated collection on discrete choice experiments in tourism. *Annals of Tourism Research, 87*, 103137. https://doi.org/10.1016/j. annals.2020.103137

Kløjgaard, M. E., Bech, M., & Søgaard, R. (2012). Designing a stated choice experiment: The value of a qualitative process. *Journal of Choice Modelling, 5*(2), 1–18. https://doi. org/10.1016/S1755-5345(13)70050-2

Kwon, M., & Remøy, H. (2019). Office employee satisfaction: The influence of design factors on psychological user satisfaction. *Facilities*, *38*(1/2), 1–19. https://doi.org/10. 1108/F-03-2019-0041

Louviere, J., Hensher, D., & Swait, J. (2000). *Stated choice methods – Analysis and applications*. Cambridge University Press.

Luce, R. D. (1959). *Individual choice behavior: A theoretical analysis*. Wiley.

Maarleveld, M., Volker, L., & Van Der Voordt, T. J. (2009). Measuring employee satisfaction in new offices–the WODI toolkit. *Journal of Facilities Management*, *7*(3), 181–197. https://doi.org/10.1108/14725960910971469

McFadden, D. (1974). Conditional logit analysis and subjective probability. In P. Zarembda (Ed.), *Frontiers in econometrics* (pp. 105–142). Academic Press.

McFadden, D. (2000, July). *Disaggregate behavioral travel demand's RUM side: A 30-year retrospective* [Paper presentation]. The International Association for Travel Behavior (IATBR) Conference, Gold Coast, Australia.

Mele, V., Bellé, N., & Cucciniello, M. (2021). Thanks, but no thanks: Preferences towards teleworking colleagues in public organizations. *Journal of Public Administration Research and Theory*, *31*(4), 790–805. https://doi.org/10.1093/jopart/muab012

Ng, W. K. B., Appel-Meulenbroek, H. A. J. A., Cloodt, M. M. A. H., & Arentze, T. A. (2019, July 3–6). *The location choice of technology-based firms: A stated choice experiment of science park alternatives* [Paper presentation]. 26th Annual Conference of the European Real Estate Society (ERES2019), Cergy-Pontoise Cedex, France.

Randle, M., Kemperman, A., & Dolnicar, S. (2019). Making cause-related corporate social responsibility (CSR) count in holiday accommodation choice. *Tourism Management*, *75*, 66–77. https://doi.org/10.1016/j.tourman.2019.05.002

Rolfö, L. V. (2018). Relocation to an activity-based flexible office–Design processes and outcomes. *Applied Ergonomics*, *73*, 141–150. https://doi.org/10.1016/j.apergo.2018.05.017

Rose, J. M., & Bliemer, M. C. J. (2009). Constructing efficient stated choice experimental designs. *Transport Reviews*, *29*(5), 587–617. https://doi.org/10.1080/01441640902827623

Rouvinen, S., & Matero, J. (2013). Stated preferences of Finnish private homeowners for residential heating systems: A discrete choice experiment. *Biomass and Bioenergy*, *57*, 22–32. https://doi.org/10.1016/j.biombioe.2012.10.010

Thurstone, L. L. (1927). A law of comparative judgment. *Psychological Review*, *34*(4), 273–286. https://doi.org/10.1037/h0070288

Tiellemans, N., Kemperman, A., Maussen, S., & Arentze, T. (2021). The influence of group decision-making on residents' preferences for sustainable energy measures of dwellings. *Building Research & Information*, *50*(4), 410–423. https://doi.org/10.1080/09613218. 2021.1992261

Train, K. (2003). *Discrete choice methods with simulation*. Cambridge University Press.

Van Cranenburgh, S., Rose, J. M., & Chorus, C. (2017, April 3–5). *On the robustness of efficient experimental designs* [Paper presentation]. International Choice Modelling Conference, Cape Town, South Africa.

Van den Berg, J., Appel-Meulenbroek, R., Kemperman, A., & Sotthewes, M. (2020). Knowledge workers' stated preferences for important characteristics of activity-based workspaces. *Building Research & Information*, *48*(7), 703–718. https://doi.org/10.1080/ 09613218.2020.1726169

Van Dongen, R. P., & Timmermans, H. J. P. (2019). Preference for different urban greenscape designs: A choice experiment using virtual environments. *Urban Forestry & Urban Greening*, *44*, 126435. https://doi.org/10.1016/j.ufug.2019.126435

Wagner, A., Gossauer, E., Moosmann, C., Gropp, T., & Leonhart, R. (2007). Thermal comfort and workplace occupant satisfaction—Results of field studies in German low energy office buildings. *Energy and Buildings*, *39*(7), 758–769. https://doi.org/10.1016/j.enbuild.2007.02.013

Wang, D., & Li, S. M. (2006). Socio-economic differentials and stated housing preferences in Guangzhou. China. *Habitat International*, *30*(2), 305–326. https://doi.org/10.1016/j.habitatint.2004.02.009

Weijs-Perrée, M., Appel-Meulenbroek, R., & Arentze, T. (2020). Analysing knowledge sharing behaviour in business centres: A Mixed Multinomial Logit Model. *Knowledge Management Research and Practice*, *18*(3), 323–335. https://doi.org/10.1080/14778238.2019.1664269

9 Delphi method

Reaching consensus on workplace performance

Chiara Tagliaro

Politecnico di Milano, Italy

9.1 Background

Delphi is a tool for negotiation through a 'structured communication process'. It is not a new method, but its application in workplace research and management is far from established. This chapter presents the method and argues that workplace studies would benefit from adopting it, especially in the new era (see the Preface of this handbook).

Delphi pursues the goal to reach stability in opinions (not necessarily consensus) on an issue that can range from the qualitative identification of factors entailing a certain event to the extraction of quantitative or semi-quantitative data in a knowledge area where no data exist, yet.

> [Delphi] is intended for use in judgment and forecasting situations in which pure model-based statistical methods are not practical or possible because of the lack of appropriate historical/economic/technical data and thus where some form of human judgmental input is necessary.
>
> (Rowe & Wright, 1999, p. 354)

The method is based on the principle that a group of people can reach a 'better' decision through an effective communication process than any single member acting alone (Linstone & Turoff, 2011). Delphi is contemplated among expert-based futures methods (Marchais-Roubelat & Roubelat, 2011) and Group Decision Support Systems (Linstone & Turoff, 2011), and it is rooted in the general theory of consistency (Kuusi, 1999). Its name definitely relates to the Delphic Oracle, being at stake the search for knowledge that is not available by other rational means through some sort of ritual (Linstone & Turoff, 2002; Marchais-Roubelat & Roubelat, 2011).

Even though Delphi can be defined as a popular and relatively well-established approach, it is not exempt from criticism. Harsh claims on the method, considering it undefined, unreliable, and poor in scientific reliability (e.g. Sackman, 1975), hindered its broad adoption until the last decade of the past century. Most likely because it misses a strict scientific procedure, sceptics of Delphi criticise it for being more of a political than a research tool (Goldschmidt, 1975). Supporters, on

DOI: 10.1201/9781003289845-9

the contrary, acknowledge its value to approach areas that are challenging to inquire with traditional scientific methods, last but not least futures analysis (Helmer, 1967; Kuusi, 1999).

One can find Delphi described from time to time as a 'study', a 'method', a 'research', a 'process', a 'methodology', an 'approach', a 'technique', a 'survey', a 'concept', an 'application', an 'inquiry', a 'panel', a 'consultation', an 'investigation', and more (Mullen, 2003). Linstone and Turoff (2002, p. 3) say that "in its design and use Delphi is more of an art than a science". There is no one 'true' Delphi. On the contrary, over-prescription is considered dangerous in Delphi applications because it will narrow its scope and inhibit its versatility (Mullen, 2003).

The Delphi process develops via an alternation of anonymous and deliberate expressions of opinions by an 'expert' panel. Usually, this entails an iterative process of sending a questionnaire to a few experts, collating the responses often accompanied by extended explanations or justifications, and resending the questionnaire, which is the same or a revised version of the original one. The second and following questionnaire submissions are frequently supplemented by a summary and elaboration of the previous responses so that the participant can reconsider his/her opinion according to the others' and, if needed, revise his/her answers based on the insights garnered from other experts.

The first record of the Delphi concept dates back to 1951 when an Air Force-sponsored RAND Corporation study employed this technique to apply expert opinion "to the selection, from the point of view of a Soviet strategic planner, of an optimal U.S. industrial target system and to the estimation of the number of A-bombs required to reduce the munitions output by a prescribed amount" (Dalkey & Helmer, 1962, p. 1). Apart from its original application in defence research, Delphi has been adopted since the 1950s for technological forecasting. Through the 1960s and 1970s, it has found fortune in healthcare and medical applications, including nursing (Mullen, 2003). Many early studies applied this technique for long-term forecasting. However, with the publication of Linstone and Turoff's (1975) edited book on Delphi, a wider audience became aware of it and started employing it in multiple domains and for different purposes (Rowe & Wright, 2011). Contributions employing a Delphi approach flourished especially throughout the 1990s. About ten years ago, the journal *Technological Forecasting and Social Change* dedicated a special issue to the Delphi technique. More recently, Delphi has also been applied in the construction and real estate sector and has spread across different geographical areas. Some of the latest studies in this field adopt Delphi to detect critical success factors of urban renewal projects in China (Chen et al., 2022); to assess the modern architectural heritage in India (Gayen et al., 2022); to determine influential indicators for office real estate price modelling in Nigeria (Yakub et al., 2022); and to develop a design quality indicators toolkit for campus facilities (Hassanain et al., 2022). It is evident that the method has had many variations to adapt to different circumstances and aims (Puglisi, 2001). In the construction and real estate sector, this technique is mostly used to create consensus on relevant variables that are likely to unfold their impact sometime in the future.

Delphi has already been used in multiple cases for what concerns the office environment. Becker and his collaborators employ Delphi in the development of the ORBIT[1] rating process (Becker, 1990). Hinks and McNay (1999) adopt the same type of study to extrapolate a bespoke set of key indicators (KPIs) for facilities management performance assessment. They describe the method as a consultative research technique, designed to merge different opinions where there is a lack of agreed knowledge. The method was also used to discover and describe new trends and their implications for the future of the office by combining the human resource, information technology, corporate real estate, and facility management perspectives and setting a common agenda of issues to address in the future workplace (De Bruyne & Gerritse, 2018). More recently, a virtual survey employing a Delphi method was launched by the International Facility Management Association (IFMA) through their latest Survey "The Experts' Assessment: New Ways of Working (NWOW) towards 2030".[2] The survey is in a real-time Delphi format that enables the respondents not only to answer questions – mostly based on Likert scales, but also to explain the reasoning behind their answers, and compare the responses to other experts instantaneously.

9.2 Argument

Futures studies techniques have been gaining increasing attention among scholars and industry professionals in the workplace. For instance, the Royal Society for the encouragement of Arts, Manufactures, and Commerce (RSA, 2019) and Johnson Controls (Ratcliffe & Saurin, 2008) have recently sponsored the employment of futures techniques to project and imagine likely workplace scenarios for the years to come. Among future studies, though, the application of Delphi in the workplace realm is still scattered besides the examples mentioned above. There are several reasons why it would make sense to adopt this method more extensively. Linstone and Turoff (1975, p. 4), in their seminal book on the Delphi method, state that the need for employing a Delphi emerges when:

- *The problem does not lend itself to precise analytical techniques but can benefit from subjective judgements on a collective basis.*
- *The individuals needed to contribute to the examination of a broad or complex problem have no history of adequate communication and may represent diverse backgrounds with respect to experience or expertise.*
- *More individuals are needed that can effectively interact in a face-to-face exchange.*
- *Time and cost make frequent group meetings infeasible.*
- *A supplemental group communication process can increase the efficiency of face-to-face meetings.*
- *Disagreements among individuals are so severe or politically unpalatable that the communication process must be referred and/or anonymity assured.*

- *The heterogeneity of the participants must be preserved to assure validity of the results, i.e. avoidance of domination by quantity or by strength of personality ('bandwagon effect').*

In fact, many of these elements are present in contemporary workplaces. Understanding workplace matters, especially in times of deep uncertainty and change as the present ones, is (1) becoming increasingly complex, (2) involving a wider range of stakeholders in 'direct-democracy' mechanisms, (3) developing into a cross-disciplinary area of inquiry, and (4) requiring the ability to project future scenarios (De Bruyne & Gerritse, 2018).

1   What seems to strongly differentiate today's work and work processes from the past is the nature, extent, and speed of change (De Bruyne & Gerritse, 2018; Kaplan & Aronoff, 1996; McGregor, 2000), which makes the world of work increasingly flexible and complex to manage. Delphi was described by Linstone and Turoff (1975, p. 3) as a method for structuring group communication so that "the process is effective in allowing a group of individuals, as a whole, to deal with a complex problem". The difficulty of managing an overwhelming knowledge load progressively led to delegation and separation of the roles.

> When organizations were small, the head of the organization was able to know virtually all the information needed to run the business. As organizations grew and became more complex, a single individual could not carry the entire knowledge load. Managers and specialists were added to the organization [...].
>
> (Kaplan & Aronoff, 1996, p. 9)

This dynamic has probably contributed to the dispersion affecting today's knowledge across different corporate departments. Therefore, engaging a growing number of actors in workplace change and related processes is necessary.

2   Workplace data are generally retained by different departments, each with its focus and lexicon (Jordan et al., 2009) according to their respective stakes, and tend to be kept separate. Chief Executive Officers (CEOs) look at growth and competitive advantage, checking for incremental revenue, speed-to-market, etc. Chief Financial Officers (CFOs) control costs and financial impacts, observing Return On Investment (ROI), occupancy costs, lease flexibility, and so on. Chief Operating Officers (COOs) are in charge of operational efficiency, therefore look at schedules, work order productivity, and space utilization. Facility Managers (FMs) deal with the building, its physical features, and related complaints. Human Resources Managers (HRMs) take care of workers' attitudes, retention and attraction, absenteeism, medical costs, and so on. All these data are rarely cross-checked among those responsible for data gathering, nor "shared widely within organizations and almost certainly not evaluated against the kinds of spaces workers occupy" (World Green Building Council, 2014, p. 62). Surveys show that only 40% of Corporate Real Estate Managers (CREM) collaborate with other business functions (i.e. HR, IT, and finance) on a regular basis (Bouri

et al., 2008). On the contrary, integration between various areas of the business should be encouraged (Amaratunga & Baldry, 2003). Moreover, workplace users include diverse individuals sharing space, time, and activities while expressing multiple and often divergent needs, among whom are employees, colleagues, customers, visitors, other stakeholders, and the company's brand (Puybaraud, 2017). It becomes clear that all these separate needs must be considered when trying to assess the efficacy of certain workplace solutions. In this context, Delphi may be the right tool to bring together, compare and contrast multiple stakes, and encourage a convergence of opinions that can be expected through the application of this approach (Helmer, 1967).

3   With the multiplication of stakes at play, workplace research comes from a broad range of disciplines, such as architecture, real estate, sociology, psychology, education, and others, that would benefit from being integrated (Kämpf-Dern & Konkol, 2017). Workplace professionals, as well, are required to master a range of competencies that belong either to ergonomics, environmental psychology, facility management, logistics, engineering, sustainability, and indoor environments. These capabilities are ideally all integrated into the background knowledge of the expertise in corporate real estate (Appel-Meulenbroek, 2014). The various fields herewith implicated complicate the spectrum and have discouraged research from an interdisciplinary strand (Fleming, 2004) and practice from establishing collaborative relationships (Duffy, 2000). That is why, the application of the Delphi method is suitable to support the development of a holistic view of the workplace (e.g. Fleming, 2004; Lavy et al., 2010; Haworth, 2016) through the merge of complementary perspectives, such as those carried by different workplace 'experts' that can cross-pollinate each other's experiences by being all involved in the iterative Delphi process.

4   Finally, the volatility, uncertainty, complexity, and ambiguity of the work environment, and of the real estate market more broadly, call for professionals that can look further in time to endorse wanted futures and avoid unwanted ones (Karjalainen et al., 2022; Saurin et al., 2008; Toivonen, 2021). Especially in the wake of the COVID-19 pandemic, most of the assumptions that seemed established in the workplace realm have been strongly challenged. Part of the effort now should be to acknowledge what new workplace factors mainly affect businesses besides the more traditional physical and human dimensions, including the virtual dimension of work. This approach requires the ability to foresee future scenarios and create visions that anticipate the coming trends and new habits of the workforce. This cannot be possible via methods other than futures methods and decision support systems.

With all that said, applying a Delphi method in workplace research and practice has the potential to involve a range of ideas and to identify a shared view on what kind of future the world of work can expect and how the related space should be managed to adapt to it progressively. In the following section, an example will be shown, demonstrating Delphi's suitability for such scopes.

9.3   Example of application/use

The experience presented in this section was carried out as part of my PhD research (Tagliaro, 2018). The study's overarching aim was to bring together multiple perspectives on workplace impacts on people and organizations. Four companies collaborated in the study using it as an advisory work to enhance their respective workplace strategies. Through the application of Delphi, the goal was to elaborate a framework of Key Performance Indicators (KPIs) suitable to monitor the extent to which workplaces were impacting organizations' needs and, ultimately, for supporting workplace management. In this case, Delphi was adopted primarily to bring workplace 'experts' to agree on meaningful KPIs for improving current workplace management practices and anticipation of future workplace strategies.

Organizations have realized that their development strongly depends on the building they occupy (Lavy et al., 2010) and the workplace is an important part of their identity (De Bruyne & Gerritse, 2018). Nevertheless, corporate leaders often underestimate their workplaces' strategic potential and real estate assets (Duffy, 2000; Lindholm & Leväinen, 2006). In addition, they miss the right tools to assess those impacts; therefore, information, when existing, is lost or remains underutilized. 52% of real estate executives is reported saying that "the lack of data and analytics to measure value and generate insights holds them back from enhancing strategic value-add to their organizations" (JLL, 2017). The current state of the practice is affected by several other drawbacks: traditional real estate metrics are considered inadequate across a complex real estate portfolio; stakeholders are demanding, but often not familiar with metrics and results; and executives have to choose among a large variety of data points to gauge performance (JLL, 2017). Also, there is still a need to go beyond mere quantitative data and identify value-adding elements for organizations.

Based on these premises, the research and consultancy experience herewith described employed Delphi to develop a framework of indicators, meaningful to a wide group of workplace users and useful to control the impact and value of the workplace. In fact, impact evaluation is common to many different fields (e.g. policy-making) where Delphi has been widely adopted to inform judgement and decisions.

In this specific application, Delphi follows the form of a *'Delphi Exercise'* (Linstone & Turoff, 1975, p. 5). The process theoretically covers four phases: (1) exploration of the subject; (2) achievement of an understanding of how the group views the issue; (3) when there is significant disagreement, evaluation of the reasons for different opinions; and (4) synthesis of the analyses and considerations. In this case, the four phases have been contracted into three stages for easiness of employment, by taking inspiration from Hinks and McNay (1999), namely: an exploration of the subject (1) was performed with one first questionnaire; focus groups or interviews addressed both the discussion on group view (2) and the reasons for different opinions (3); and one final questionnaire was employed for synthesis (4). This particular application took inspiration from the EFTE (Estimate, Feedback, Talk, Estimate) approach, which includes open debate phases in different rounds (see Puglisi, 2001). Below these different stages will be described in detail.

9.3.1 *The setting of the study*

Before starting the Delphi, some preparation was necessary, on the one hand, to identify the panel of experts and to become familiar with them and, on the other hand, to organize the KPIs to be discussed throughout the process.

1 *Preparation.* This phase entailed KPI scouting, analysis, and systematization. A broad literature review was conducted to match indicators from the scientific literature with indicators from evaluation and rating tools. Finally, a list of 169 items was compiled (after appropriate selection and cleaning of a broader list). These were organized into nine thematic categories (i.e. environmental quality, building operation and management, space usage, business effectiveness, costs, value/return/yield, productivity/ways of working, user attitude, and staff characteristics) and three classes of impact/value (i.e. organizational, environmental, and social) – see Table 9.1. Despite work to avoid redundancy, the long list still contained a few similar indicators. This was functional to stimulate the group discussion and come up with the most suitable indicator among similar choices.

2 *Selection of the case companies.* The experts were recruited from four different companies identified by convenience sampling with the following criteria:

- Medium-large sized companies (500 employees or more), headquartered in Italy. This was functional to guarantee a reasonable complexity of the organizational and managerial structure along with similar cultural background, and to avoid language issues.
- Industry diversity within knowledge-based organizations. A pharmaceutical, a utility, an automation technology, and an insurance company were invited in order to assure different approaches to business problems.
- Companies that have renewed their premises or moved to new premises in recent times (less than five years) or are going to do it within the next five years. This to make sure that they were sufficiently familiar with the new ways of working and demonstrated an interest in aligning their workplace with organization needs. In the end, the sample of four companies comprised two that had already gone through a workplace change and two that, at the moment of the study, were preparing for an imminent change.

Table 9.1 Categories of indicators by classes of impact

Class – Impact – Value		
Economic-financial/organizational	*Environmental*	*Social*
Cost	Building operation and management	Productivity/Ways of Working
Value/Return/Yield	Environmental quality	User attitude
Business effectiveness	Space usage	Staff characteristics

3 *Getting familiar with the sample.* Before Delphi administration, the researcher got acquainted with each company's culture and experience with workplace management and change. This step was undertaken through semi-structured interviews with a few representatives in each company. These encounters explored the current workplace strategies and verified the degree of familiarity with performance assessment techniques at a corporate level. Besides, these dialogues were used to grasp suggestions for innovative KPIs and potentially expand the 169-item list. Finally, the best modalities to perform the Delphi research in each organization were arranged through these meetings.

4 *Identification of the expert panel.* Experts in this case were considered all the people who have a direct experience of the workplace in the different organizations and who have a different level of engagement with the management process. Ten user categories were identified in a previous research phase: CEO/owner/president, financial admin, CREM, FM, HRM, Engineering and space planning, IT, executives and managers, employees, and consultants/collaborators/interns. At least one representative of each category was to be included in the expert group. Their nomination happened thanks to direct contacts in the companies. A total of 40 participants composed the initial group, though some people progressively dropped participation during the process (Table 9.2). The following Delphi stages were performed separately in each company, namely in four groups of (about) ten people each.

9.3.2 Exploration of the subject

A first online questionnaire was prepared and administered to all 40 participants to get accustomed to the subject and start making their choices. The participants received the list of KPIs divided into the nine thematic categories and were asked to anonymously evaluate the level of importance of each indicator by stating whether they thought it was 'essential', only 'desirable', or of 'tertiary level importance'.

Table 9.2 Number of participants in the subsequent Delphi stages by expert categories

	Questionnaire 1	Interviews/ Focus groups	Questionnaire 2
Expert categories	**Total**	**Total**	**Total**
Consultant/Collaborator/Internee	5	4	4
Corporate Real Estate Management	4	4	2
Designer/architect/engineer	4	3	2
Employee/clerical in other department	4	3	3
Facility Management	3	2	1
Human Resources Management	4	4	3
ICT Department	3	3	4
Manager in other department	5	4	3
Office Financial Administration	3	3	2
Owner/President/CEO	4	3	1
Total number of participants	**39**	**33**	**25**

Table 9.3 KPIs in the initial list vs. KPIs after Questionnaire 1

KPI category	KPIs in the initial list		KPIs in the list after Questionnaire 1		
	Number	%	Number	%	Sum of votes
Building operation and management	34	20%	11	26%	314
Environmental quality	9	5%	9	21%	299
Productivity/Ways of Working	34	21%	5	12%	126
User attitude	16	9%	5	12%	123
Staff characteristics	13	8%	5	12%	113
Space usage	10	6%	4	9%	87
Business effectiveness	18	11%	3	7%	71
Costs	24	14%	1	2%	29
Value/Return/Yield	10	6%	0	0%	0
Total	**169**	**100%**	**43**	**100%**	**1162**

Moreover, the opportunity to state "I do not know, I do not understand this indicator" was also given, since participants may have been not familiar with some technical indicators extraneous to their own field of competence. The option to add indicators to each of the nine categories, or independently from the proposed categories at the end of the questionnaire, was also given if participants considered that some important aspects were missing.

Once all the participants completed the questionnaire, the indicators were put together into an ordered list based on the frequency of essential votes. To accomplish the required elaborations, indicators were assigned a numerical value based on the rating they received (Essential – 3; Desirable – 2; Of tertiary importance – 1; I do not know – 0). A list of 43 indicators collected all those rated as 'essential' by more than the 50% of the sample, which meant those attracting more than 20 essential votes (Table 9.3). This list was brought forth to the next steps.

This initial step of the process allowed to:

a Reduce the list of KPIs to a manageable number of items based on the convergence of opinion (expressed anonymously) across multiple workplace users;
b Identifying the user categories who did not understand some of the indicators, and question whether this misalignment was important to fix or should be considered 'physiological'; and
c Finding out convergence/divergence of opinions across users, and across different companies to argue in the following step.

9.3.3 Acknowledgement of group views and different opinions on essential KPIs

The second stage of the Delphi method aimed to compare, contrast, and discuss openly among the expert group the choices that were made anonymously in the previous activity. It is worth pointing out that, at this point, the participants had all already gone through the previous questionnaire individually, and therefore the assumption was that they had become accustomed to the topic and had formed some conscious ideas. The discussion was guided to address (i) the definition of each indicator, which should be comprehensible to all the users; (ii) the placement of each indicator in the corresponding category and impact class; and (iii) the actual importance of each indicator compared to the others – whether it should be dropped or kept in the list.

Given that six indicators had been added while answering Questionnaire 1, the list of 43 indicators was integrated with those six before bringing it to group discussion. Therefore, 49 indicators (43 + 6) were finally debated, also considering the similarity of the newly proposed KPIs with already existing ones. In this phase, people could still erase unnecessary indicators, retrieve indicators from the initial 169-item list, or suggest new indicators, which they believed relevant to all workplace users.

The discussion was undertaken through either one-to-one interviews or focus groups. The latter were encouraged by the researcher, but not all the companies had the possibility to organize them. Therefore, single interviews were scheduled when necessary and the researcher put together the results of the individual conversations afterwards, through qualitative analysis. Thanks to this process, the KPI list was compressed from 49 to 33 indicators. Thanks to group discussion, each item was confirmed or re-assigned into a thematic category and a class of impact/value (see Tables 9.4 and 9.5). Many suggestions stood up about the opportunity of having a dashboard where KPIs could be compared to internal benchmarks. For refinement of the final set of indicators, a few people proposed to organize indicators in a tree-like structure. In fact, a passage encompassing KPI ordering was faced in the next step of analysis, through Questionnaire 2.

This second step of the process allowed to:

a Discuss workplace impacts/value openly, thus fostering opinion sharing among users who rarely can meet and exchange their respective ideas. This greatly enriched the view over impacts to recognize that organizational, environmental, or social impact cannot be evaluated in isolation but are transversal. However, participants all agreed upon the usefulness of classification and categorization, especially for easier recognition of the professionals dealing with the related indicators.

b Come up with new creative ideas on how to compose and represent the final set of indicators, in order for it to become a dashboard for information sharing among all workplace users. A few people proposed to give evidence to prioritization. Questionnaire 2 was set up exactly to let prioritization and applicability of KPIs emerge.

9.3.4 Synthesis and future strategies

At this point of the Delphi process, the short list of 33 indicators, appropriately reordered and systematized by the researcher as per it was re-elaborated during focus groups and interviews, was submitted once again to the same participants.

Table 9.4 Building operation and management KPIs before discussion

	Class – Impact/Value		
Category	*Financial/ organizational*	*Environmental*	*Social*
Building operation and management		• Accessibility for disabled • Ethics, health and safety practices (Health and safety, Provision of safe environment) • Networking IT • Competence of staff • Reliability • Effectiveness of help desk service, Response time, Responsiveness to problems • Telecommunications • Standards of cleaning • Security • Correction of faults • Resource consumption (energy, water, materials), Sustainability objectives (waste, energy consumption, etc.), Environmental sustainability of buildings - *Number of requests made vs. number or requests met with timing* - *Number of audits/month* - *Response time* - *Resolution time* - *Cost of corrective intervention*	

Legend: The dotted list contains KPIs from the literature; the dashed list collects the KPIs added in Questionnaire 1.

Table 9.5 Building operation and management KPIs after discussion

	Class – Impact/Value		
Category	*Financial/ organizational*	*Environmental*	*Social*
Building operation and management	• Cost of corrective intervention	• Standards of cleaning • Number of FM requests made vs. number or requests met with timing • Resource consumption (energy, water, materials), Sustainability objectives (waste, energy consumption, etc.), Environmental sustainability of buildings	• Design for All (Accessibility for disabled) • Quality of communication strategies to encourage ethics, health and safety practices • Reliability of the maintenance service (Competence of facility management staff)

Table 9.6 Average grade by KPI categories and users

Overall averages	EQ	BOM	PW	SU	CO	UA	SC	BE	VRY	Total
CREM	5,61	5,14	5,00	5,50	5,50	4,67	4,67	5,00	4,50	**5,07**
Designer	5,83	5,86	5,92	7,00	5,00	5,00	4,33	6,75	3,00	**5,41**
FM	3,11	3,86	4,00	7,00	5,00	6,33	7,00	0,00	0,00	**4,03**
Owner	4,33	4,86	4,50	4,00	6,00	3,00	2,67	2,50	0,00	**3,54**
Consultant	5,91	6,10	5,97	5,00	5,75	5,42	5,39	4,67	5,00	**5,47**
Manager	6,64	6,10	5,97	5,67	5,60	5,83	5,77	5,20	5,50	**5,81**
HRM	6,35	6,48	6,28	6,00	6,00	6,00	6,28	5,33	5,00	**5,97**
CFO	5,50	6,07	6,25	6,00	4,00	6,17	5,83	5,75	6,00	**5,73**
Employee	6,54	5,86	5,89	4,67	4,67	5,56	6,11	4,67	5,00	**5,44**
ICT	6,17	5,49	5,19	6,00	6,33	5,14	5,39	3,50	3,50	**5,19**
Total	**6,09**	**5,88**	**5,80**	**5,79**	**5,57**	**5,54**	**5,52**	**5,15**	**4,87**	**5,58**

Legend: EQ = Environmental Quality; UA = User Attitude; BE = Business Effectiveness; SC = Staff Characteristics; PW = Productivity/Ways of Working; BOM = Building Operation and Management; CO = Costs; SU = Space Usage; VRY = Value/Return/Yield.

Submission happened through a second online questionnaire to fill up anonymously, where people were asked to (i) rate the level of importance they attributed to each KPI (on a scale from 1 to 7); and (ii) declare whether each indicator was currently measured in their company or not, based on their personal knowledge.

Matching this information allowed the assessment of indicators' applicability to workplace management. At first, participants were requested to assign a 'priority' grade to each KPI from 1-low importance to 7-high importance. It was possible to avoid awarding a KPI if considered erasable from the final list. Priority grade corresponded to the sum of all grades received by all users who participated in the questionnaire. The grades different users assigned to KPIs were on average quite flattened across user categories, except a few cases, and partially resembled the results obtained with Questionnaire 1. This showed that the choices made by all user categories were quite stable at this point and confirmed the convergence of opinion among them (Table 9.6). Moreover, the priority grade indicated the importance workplace users attributed to each indicator, which reflected people willingness to contribute to KPI measurement, their level of engagement in information sharing, and the value they credited to that specific aspect of the workplace (by category and class).

Second, people were asked about 'measurement practices' in their respective organizations, namely if each KPI was already gathered in their company (possible answers "Yes/No/I don't know"), as a proxy to indicate 'readiness for adoption'. The probability of data availability was obtained through the mean average of *Yes* answers, compared to the theoretically possible 100% of Yes answers. The assumption was that, if data were already available in most of the companies, this corresponded to relative preparedness of technologies, methods, and competences to bring the KPI from theory to operation.

By representing these two elements on a dispersion graph, it was possible to obtain an 'opportunity scale' for KPI management (Figure 9.1). Sequential numbering

Legend: ■ Environmental/spatial impact

★ Social/perceptual impact

● Financial/organizational impact

4. Design for All
14. Psychophysical wellness
15. Multiculturality (staff composition)
16. Acoustics comfort
18. Adequacy of space, suitability of premises and functional environment

1. Accessibility to ICT networks
2. Indoor air quality and ventilation
3. Number of FM requests made vs.
    number of requests met with timing

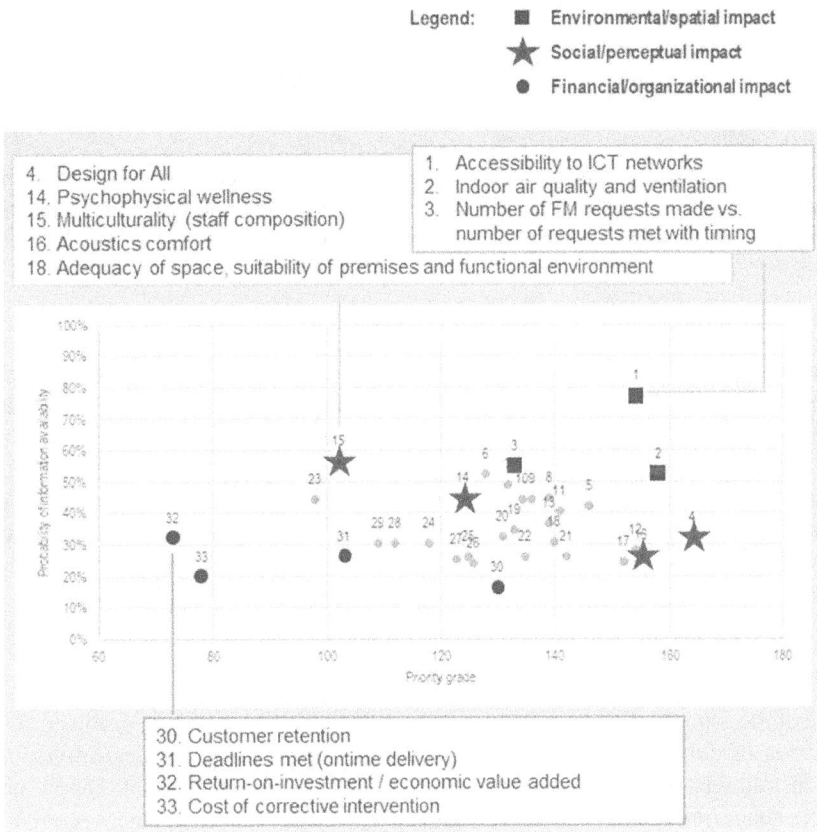

30. Customer retention
31. Deadlines met (ontime delivery)
32. Return-on-investment / economic value added
33. Cost of corrective intervention

Figure 9.1 Opportunity for KPI measurement and management

of KPIs in the dispersion graph is based on the sum of the percentage level of priority grade on the maximum priority obtainable and the percentage number of *Yes* answers on the total number of participants. Indicators that fall in the top right area are those scoring a high priority grade and a high probability that data are already available to populate that KPI.

This third and last step of the process allowed us to:

a Produce an operational tool in a matrix-like structure that can help all workplace users interpret indicators and their interrelations and, in turn, support the elaboration of strategies throughout the design, management, and use phases of the workplace;

b Attribute the 'ownership' for KPI monitoring, namely identify the roles entitled to control indicators in the financial/organizational class (typically CFO, Owner, CREM, and FM); in the environmental class (i.e. FM, CREM, and ICT); and in the social class (i.e. HR, FM, CREM;

c Recognize the most 'active' user categories and those who, instead, are not yet involved enough in the workplace-making processes (in the case under analysis, for instance, HR and ICT seemed rather detached from the topic) and start elaborating on the opportunity to attract them to workplace matters better; and finally
d Realize that when companies go through a workplace change, they tend to become keener on collaborating into cross-departmental and cross-role activities towards the achievement of common interest, which Delphi can greatly support.

9.4 Implications

In the case presented above, Delphi proved effective in bringing together several workplace users, who usually remain isolated in the organizational debate around workplace matters, and make them focus on a common goal (e.g. the systematization of KPIs to support workplace management strategies). The method has both research and practical implications. On the one hand, it may help researchers answers questions of scientific relevance and advance academic knowledge. On the other hand, due to its very nature, it implies deep interactions among the participants, including the same researcher, and thus becomes an operative tool.

9.4.1 Method relevance to research

One of the aspects of Delphi that generates mixed feelings lies in its intrinsic combination of quantitative and qualitative approaches, by merging a positivist and a constructivist attitude to research (Mullen, 2003). To some extent, Delphi produces quantified results within a recognizably positivist tradition, whereas, to some other extent, the definition of the problem and the solutions to it, produced by those who are the same subjects of the research, place it close to a constructivist position. Delphi straddles the divide between qualitative and quantitative methodologies (Critcher & Gladstone, 1998). Despite this appreciation, still many recent criticisms of Delphi and attempts to prescribe the 'correct' process stem from the positivist critique. This ambivalence can, indeed, be tricky during research, and somehow difficult to handle. Nevertheless, it also demonstrates the uniqueness of the approach and places it in a very interesting position in-between disciplines. This might be the secret of its success across multiple disciplinary areas, from the hard to the soft sciences, since its inception some 70 years ago. Thus, it is important that Delphi is not confined to limited application areas but exploits the potential of cross-fertilization across disciplines (Mullen, 2003).

Among other classic targets of criticism, sources of controversy and misunderstandings about Delphi's relevance to research are: the use of an 'expert' panel, the claim of consensus, questionnaire construction, and the alternation of anonymity and interaction between panel members (Mullen, 2003). These are all partial limitations embedded in the method that can be as well considered strengths and opportunities for deeper and multi-faceted interpretation of the phenomenon under analysis. For instance, the alternation of individual reflection and group discussion is a great way to let people think on their own when their single characteristics (either professional or psychological) can emerge, and balance those reflections with group discussion when other opinions may come up and smoothen or radically change some of the initial ideas. This openness is rather unique to the Delphi approach and might be of particular benefit in the workplace realm where research struggles to keep together so many different stakeholders needs and managerial strategies. Nonetheless, caution is recommended especially in corporate environments where open expression of opinions may lead to repercussions on the individuals or unpleasant re-adjustment of power relations. The flexibility of the method though makes it possible to opt for overt or more covered idea sharing. In addition, the review of opinions is encouraged during the process, so that the same researcher responsible for data elaboration in due course goes through some sort of peer-to-peer confrontation that usually happens only in a later stage of the research process. Such an exchange puts Delphi in a privileged position in-between research and practice.

Finally, many suggest that Delphi should be used in combination with other methods as a part of a wider process (Rowe & Wright, 2011). In order to encourage broader adoption of the method, studies could do a better job of describing the application of the process itself, by giving details on the scoring, aggregation, and feedback of methods employed (Mullen & Spurgeon, 2000), as I did in this chapter and few other reports do.

9.4.2 *Method relevance to practice*

We are in a new era when presence in the office is unpredictable and engagement is difficult to assure due to opportunities for conventional meetings thinning out. Delphi might foster a constructive debate about the workplace and create common ground among people who share similar stakes in the office environment but do not live it and use it in the same moments nor in similar ways. Delphi comes handy to compare and contrast views of the workplace that are unlikely to emerge otherwise. Moreover, compared to conventional surveys, Delphi allows some degree of interaction and exchange, via feedback and justification, between respondents and between respondents and researchers. Therefore, it has the advantage of reflecting a process of decision-making that, by its nature, happens in a progressive way by reconsidering judgments and revising conclusions. This is of utmost importance for companies and decision-makers in a time when working conditions are more changeable and uncertain than ever before. Finally, Delphi may help identify a set of agenda items to intervene in the workplace by priority and support its progressive reassessment while the future is getting closer. In the example shown above, the method supported multiple achievements, including letting different stakes emerge and express themselves in a protected and constructive environment; democratizing workplace management and putting all users on the same level (executives, employees, building operators, etc.); and helping take a direction for future management strategies in the workplace based on shared priorities. Especially when figuring out new workplace performance indicators, Delphi stimulates innovative ideas about KPI measurement and management, responding to the inadequacy of traditional real estate metrics for complex environments; making stakeholders more familiar with metrics and results; helping executives make sense of a large variety of data points to gauge performance (JLL, 2017); and going beyond mere quantitative data to identify value-adding elements for organizations.

Therefore, the method can be used to change the company culture and rebalance power relations, even though it must be considered that the method is very old and back when it was initiated the world of work was very different (e.g. hierarchy in decision-making, the need for anonymity, and more). Even more so, it depends on the working culture of each organization how the method can be implemented.

One thing to be noted is that the method was originally for the purpose of envisioning something that did not yet exist, but in workplace management, so far, it has been used more for shorter time perspective than in other real estate areas. This might be due to the fact that the time span of land use planners, institutional investors, etc., is longer than that of corporate real estate players and workspace users.

Overall, this method resembles the management techniques for stimulating creativity and innovation, which are based on the alternation of divergent (openness) and convergent (focus) phases in order to bring in new ideas and verify if they make sense. Companies may already be familiar with them and, therefore, trust the process and be more likely to participate. Making the implementation of Delphi a more common practice will be useful for the full deployment of workplaces' potential.

9.5   Conclusions

This contribution adds to other studies providing guidance for applying Delphi in the workplace realm. The chapter has introduced the approach proposed by Delphi and demonstrated its potential in workplace research and management for understanding relations between elements, forecasting and futures studies, priority setting, and user involvement. The method does not come without risks and limitations. Attention should be given to creating the necessary heterogeneity of panel participants, improving question formulation, enhancing panellist retention throughout subsequent rounds, enhancing information exchange between participants during feedback stages, and combining Delphi with other techniques (Rowe & Wright, 2011).

For instance, in the experience presented above one of the companies showed resistance at some point and many participants dropped out of interviews and, later, Questionnaire 2. A lack of bonding with the research aim should be expected. A Delphi technique is not costly in monetary terms but requires a prolonged series of interactions between participants and researchers, which is difficult to evaluate at the very beginning of the process in terms of effort for the companies over time. Despite good intentions, many factors can intervene in the meantime to obstruct companies from participation during the whole Delphi development. Nevertheless, the dropout rate may have been also due to the chosen process entailing in-person exchange. The risk is that the loudest voice rather than the soundest argument may carry the day; or a person may be reluctant to abandon a previously stated opinion in front of his peers; and when one disagrees with the boss, the discomfort might be discouraging. Anyways, Delphi gives a good possibility to change views without losing face in front of colleagues. Therefore, the study setting must be carefully evaluated to prevent uneasy situations.

There is also significant limitation about generalizability, especially when the sample size remains small – this is counterbalanced nowadays by online Delphi that can be spread widely across the population. Considering the example in this chapter, someone could contend that every building and every organization are different, so that every assessment should be done on a case-by-case basis. Indeed, the experts in this case remained isolated by company and it was the researcher's responsibility to combine their viewpoints. Delphi is often used for experts that come from different fields and organizations. In further developments of this study, some outsiders could be invited, or cross-company meetings could be organized to foster the generalizability of the results. Some assert that relevant working performance measures should be defined by each individual company or business function (Hinks & McNay, 1999; Kämpf-Dern & Konkol, 2017) as many criteria vary greatly, e.g. (a) type of users (typically facility manager, executive level management, etc.); (b) nature of the organization (private or public); (c) focus of the assessment (e.g. financial, functional, physical); and (d) industry trends (Lavy et al., 2014). However, when buildings are constructed for particular purposes and at a certain time in history, they tend to have many similarities, or at least they apply similar design strategies (Becker, 1990); hence, it is likely that they share a similar view towards the future.

Organizations do change over time, which is why it is important for assessment tools to treat departments, divisions, or other organizational units with sufficient detail. Indeed, the very aim of Delphi, when it is adopted for sincere research purposes and not for political reasons, should be not to force artificial consensus among the parties but to let different opinions emerge, challenge them against one another, and find stability.

Finally, two questions remain still open regarding the application of a Delphi method, especially when it applies to scenario creation (Marchais-Roubelat & Roubelat, 2011) – which is not the case of the study presented in this chapter but is worth discussing for prospective applications. First, does the knowledge of future scenarios affect the production of a self-fulfilling prophecy? Throughout the process and afterwards, the behaviour of the involved experts might be oriented so that the events themselves tend to unfold as per they were virtually forecasted. Second, research in neuroscience suggests that different areas of the brain react to different levels of uncertainty. Does this mechanism have the same effect on all experts or could this somehow influence in different ways the experts' engagement in the process? We should still consider that the purpose of future studies is not to predict a certain future but to picture alternative futures and steer the stakeholders into a different direction than the one that would naturally occur given certain conditions. As for now, it looks like Delphi in workplace contexts has been used typically for other purposes than foreseeing possible futures or other future-related issues. It would be interesting to see more experimentation on this kind of application.

This chapter should conclude with some recommendations for the employment of the method. Nevertheless, Linstone and Turoff (1975, p. 6) invite to beware of the potential problems arising "when a Delphi designed for a particular application is taken as representative of all Delphis". In sum, the method must be applied with a critical approach to enable researchers and practitioners find answers and even new questions that they did not know how to express beforehand (Rowe & Wright, 2011).

Acknowledgments

I want to thank Saija Toivonen and Lassi Tähtinen from Aalto University, School of Engineering, Department of Built Environment, who dedicated part of their time to a visiting period in Italy (in the context of RESCUE project funded by the Academy of Finland – https://www.rescue-finland.com/) to carefully reading this chapter and providing their precious feedback. Given their familiarity with the method and future studies in general, they acted as external reviewers and provided significant input for the development and refinement of this chapter.

9.7   Further Reading

About futures studies, their foundations, and epistemological considerations, it is recommended to look at the seminal books by the sociologist Wendell Bell:

Bell, W. (1996). *Foundations of futures studies: Human science for a new era, vol. 1, 'History, purposes and knowledge' and vol. 2, 'Values, objectivity and the good society'.* Transaction Publishers.

By the same author, there is a commentary on the meaning and importance of future studies for the society:

Bell, W. (1998). Making people responsible: The possible, the probable, and the preferable. *American Behavioral Scientist, 42*(3), 323–339. https://doi.org/10.1177/0002764298042003004

For those who want to have a broad and fast overview of the study of the futures, Puglisi (2001) conference paper compares and contrasts multiple different futures studies methodologies. It describes in sufficient depth some interesting classification of futures techniques and discusses in particular: forecasting methods, environmental scanning, simulation and modelling, black-view mirror analysis, Delphi, scenarios, visioning, futures biographies, futures workshops, causal layered analysis:

Puglisi, M. (2001). The study of the futures: an overview of futures studies methodologies. In D. Camarda & L. Grassini (Eds.), *Interdependency between agriculture and urbanization: Conflicts on sustainable use of soil and water* (pp. 439–463).

Digging into futures methods applied in workplace research, Saurin's (2012) doctoral thesis provides an intriguing combination of methods within the Prospective Through Scenarios process to assist organizations and facility managers in workplace planning for effective long-term strategies:

Saurin, R. (2012). *Workplace futures: A case study of an adaptive scenarios approach to establish strategies for tomorrow's workplace* [Doctoral thesis]. Technological University Dublin. https://doi.org/10.21427/D7PG6H

Finally, Marchais-Roubelat and Roubelat (2011) extensively discuss whether 'Delphi' is just a namesake for the Delphic Oracle or if it makes sense to seek a parallel between the two approaches, especially in order to better understand the characteristics of the knowledge revealed, on the one hand, and the role of the actors in the inquiring process, on the other:

Marchais-Roubelat, A., & Roubelat, F. (2011). The Delphi method as a ritual: Inquiring the Delphic Oracle. *Technological Forecasting and Social Change*, *78*(9), 1491–1499. https://doi.org/10.1016/j.techfore.2011.04.012

Notes

1 The Office Buildings and Information Technology (ORBIT) studies were carried out in two separate stages, in the 1983 and the 1985. Led by DEGW, the aim of the studies was to identify the impact of new technologies on office design.
2 The findings of this study sponsored by HOK, JLL, Planon, and Savills were presented during World Workplace in Nashville on 28–30 September 2022. https://events.ifma.org/worldworkplace/2022/conference_schedule.cfm

References

Amaratunga, D., & Baldry, D. (2003). A conceptual framework to measure facilities management performance. *Property Management*, *21*(2), 171–189.
Appel-Meulenbroek, H. A. J. A. (2014). *How to measure added value of CRE and building design: knowledge sharing in research buildings* [Doctoral thesis]. Technische Universiteit Eindhoven.
Becker, F. (1990). *The total workplace: Facilities management and the elastic organization*. Van Nostrand Reinhold.
Bouri, G., Acoba, F. J., & Wu, P. (2008). Organizational design: Leading practice. *The Leader*, *7*(4), 14–24.
Chen, Y., Han, Q., Liu, G., Wu, Y., Li, K., & Hong, J. (2022). Determining critical success factors of urban renewal projects: Multiple integrated approach. *Journal of Urban Planning and Development*, *148*(1), 04021058. https://doi.org/10.1061/%28ASCE%29UP.1943-5444.0000775
Critcher, C., & Gladstone, B. (1998). Utilizing the Delphi technique in policy discussion: A case study of a privatized utility in Britain. *Public Administration*, *76*(3), 431–450.
Dalkey, N., & Helmer, O. (1962). *An experimental application of the Delphi method to the use of experts. Memorandum RM-727/1-Abridged*. The RAND Corporation.

De Bruyne, E., & Gerritse, D. (2018). Exploring the future workplace: Results of the futures forum study. *Journal of Corporate Real Estate, 20*(3), 196–213. https://doi.org/10.1108/JCRE-09-2017-0030

Duffy, F. (2000). Design and facilities management in a time of change. *Facilities, 18*(10/11/12), 371–375.

Fleming, D. (2004). Facilities management: A behavioral approach. *Facilities, 22*(1/2), 35–43.

Gayen, P., Hajela, A., & Kumar, J. (2022). An approach for assessing the modern architectural heritage of India. In Versaci, A., Cennamo, C., & Akagawa, N. (Eds.), *Conservation of architectural heritage (CAH). Advances in science, technology & innovation* (pp. 287–300). Springer.

Goldschmidt, P. G. (1975). Scientific inquiry or political critique? *Technological Forecasting and Social Change, 7*(2), 195–213.

Hassanain, M. A., Sanni-Anibire, M. O., & Mahmoud, A. S. (2022). Development of a design quality indicators toolkit for campus facilities using the Delphi approach. *Journal of Architectural Engineering, 40*(9/10), 594–616. https://doi.org/10.1108/F-09-2021-0084

Haworth. (2016). *Enabling the organic workspace: Emerging technologies that focus on people, not just space.* https://www.thercfgroup.com/files/resources/Emerging-Technologies-that-Focus-on-People-Not-Just-Space.pdf

Helmer, O. (1967). *Analysis of the future: The Delphi method.* The RAND Corporation.

Hinks, J., & McNay, P. (1999). The creation of a management-by-variance tool for facilities management performance assessment. *Facilities, 17*(1/2), 31–53.

JLL. (2017). *Choosing metrics that matter.* https://www.futureofwork.jll/financial-performance/metrics-that-matter/

Jordan, M., McCarty, T., & Velo, B. (2009). Performance measurement in corporate real estate. *Journal of Corporate Real Estate, 11*(2), 106–114.

Kämpf-Dern, A., & Konkol, J. (2017). Performance-oriented office environments - framework for effective workspace. *Journal of Corporate Real Estate, 19*(4), 208–238.

Kaplan, A., & Aronoff, S. (1996). Productivity paradox: Worksettings for knowledge work. *Facilities, 14*(3/4), 6–14.

Karjalainen, J., Heinonen, S., & Taylor, A. (2022). Mysterious faces of hybridisation: An anticipatory approach for crisis literacy. *European Journal of Futures Research, 10*(21). https://doi.org/10.1186/s40309-022-00207-5

Kuusi, O. (1999). *Expertise in the future use if generic technologies. Epistemic and methodological considerations concerning Delphi studies.* Acta Univesitas Oeconomicae Helsingiensis, A-159, Helsinki School of Business Administration. https://core.ac.uk/download/pdf/153491932.pdf

Lavy, S., Garcia, J. A., & Dixit, M. K. (2010). Establishment of KPIs for facility performance measurement: Review of literature. *Facilities, 28*(9/10), 440–464.

Lavy, S., Garcia, J. A., & Dixit, M. K. (2014). KPI's for facility's performance assessment, part II: Identification of variables and deriving expressions for core indicators. *Facilities, 32*(5/6), 275–294.

Lindholm, A.-L., & Leväinen, K. I. (2006). A framework for identifying and measuring value added by corporate real estate. *Journal of Corporate Real Estate, 8*(1), 38–46.

Linstone, H. A., & Turoff, M. (1975). *The Delphi method. Techniques and applications.* Addison-Wesley Publishing Company.

Linstone, H. A., & Turoff, M. (2002). *The Delphi method. Techniques and applications.* https://web.njit.edu/~turoff/pubs/delphibook/delphibook.pdf

Linstone, H. A., & Turoff, M. (2011). Delphi: A brief look backward and forward. *Technological Forecasting and Social Change, 78*(9), 1712–1719. https://doi.org/10.1016/j.techfore.2010.09.011

Marchais-Roubelat, A., & Roubelat, F. (2011). The Delphi method as a ritual: Inquiring the Delphic Oracle. *Technological Forecasting and Social Change, 78*(9), 1491–1499. https://doi.org/10.1016/j.techfore.2011.04.012

McGregor, W. (2000). The future of workspace management. *Facilities, 18*(3/4), 138–143.

Mullen, P. M. (2003). Delphi: Myths and reality. *Journal of Health Organization and Management, 17*(1), 37–52. https://doi.org/10.1108/14777260310469319

Mullen, P. M., & Spurgeon, P. (2000). *Priority setting and the public*. Radcliffe Medical Press.

Puglisi, M. (2001). The study of the futures: An overview of futures studies methodologies. In D. Camarda & L. Grassini (Eds.), *Interdependency between agriculture and urbanization: Conflicts on sustainable use of soil and water* (pp. 439–463). CIHEAM.

Puybaraud, M. (2017). *Workplace powered by human experience*. Jones Lang LaSalle. https://www.jll.it/it/tendenze-e-ricerca/research/workplace-powered-by-human-experience

Ratcliffe, J., & Saurin, R. (2008). *Towards tomorrow's sustainable workplace: Imagineering a sustainable workplace future*. Johnson Controls.

Rowe, G., & Wright, G. (1999). The Delphi technique as a forecasting tool: Issues and analysis. *International Journal of Forecasting, 15*(4), 353–375.

Rowe, G., & Wright, G. (2011). The Delphi technique: Past, present, and future prospects. *Technological Forecasting and Social Change, 78*(9), 1487–1720.

RSA. (2019). *The four futures of work. Coping with uncertainty in an age of radical technologies*. https://www.thersa.org/reports/the-four-futures-of-work-coping-with-uncertainty-in-an-age-of-radical-technologies

Sackman, H. (1975). Summary evaluation of Delphi. *Policy Analysis, 1*(4), 693–718.

Saurin, R., Ratcliffe, J., & Puybaraud, M. (2008). Tomorrow's workplace: A futures approach using prospective through scenarios. *Journal of Corporate Real Estate, 10*(4), 243–261. https://doi.org/10.1108/14630010810925118

Tagliaro, C. (2018). *A place for the workplace to work: A system of performance indicators for strategic design, management and use of the workplace* [Doctoral thesis]. Politecnico di Milano.

Toivonen, S. (2021). Advancing futures thinking in the real estate field. *Journal of European Real Estate Research, 14*(1), 150–166. https://doi.org/10.1108/JERER-01-2020-0003

World Green Building Council. (2014). *Health, wellbeing & productivity in offices. The next chapter for green building*. https://builddesk.co.uk/wp-content/uploads/2015/01/Health-Wellbeing-and-Productivity-in-Offices-The-next-chapter-for-green-building-Full-Report_0.pdf

Yakub, A. A., Achu, K., Ali, H. M., & Abdul Jalil, R. (2022). An analysis of the determinants of office real estate price modelling in Nigeria: Using a Delphi approach. *Property Management, 40*(5), 758–779. https://doi.org/10.1108/PM-08-2021-0060

10 Social network analysis

Studying social interactions and relations in the workplace

Yaoyi Zhou

Virginia Tech, USA

10.1 Background

Social networks are defined as a set of nodes (or network members) that are tied by one or more types of relations (Wasserman & Faust, 1994). For those not familiar with network research, a network is a set of actors connected by a set of ties. The actors (often called "nodes") can be persons, teams, organizations, concepts, etc. Ties connect pairs of actors and can be directed (i.e., send a message to someone) or undirected (as in being physically proximate) or valued (i.e., strong vs. weak friendship). A set of ties of a given type (such as friendship ties) constitutes a binary social relation, and each relation defines a different network (e.g., the friendship network is distinct from the advice network). Different kinds of ties are typically assumed to function differently, although empirically they might be correlated. Some typical phenomena such as small-world effect (Pool & Kochen, 1978), the strength of weak ties (Granovetter, 1983), and many others may be observed in social networks.

Social Network Analysis (SNA) is a research methodology that seeks to identify underlying patterns of social relations based on the way actors are connected with each other (Breiger, 2004; Scott & Carrington, 2011). The unit of analysis in SNA is not the individual, but the relationships or interactions that occur between members of the network. Using SNA, the social environment can be mapped as patterns of relationships among interacting members.

When applying a network perspective, SNA can be used to indicate how a certain individual is connected to others, and also indicates the cohesion of a network. There are two key indicators used in SNA: "density" and "centrality." Density provides a measure of the overall "connections" between the participants. The more participants connected to one another (by, for example, their message exchanges), the higher will be the density value of the network. While centrality indicates the extent to which an individual was connected to other actors within a network (Wasserman & Faust, 1994).

In organizational studies, SNA has been widely adopted to study group dynamics, social interaction, and communications in organizations at different levels. Borgatti and Foster (2003) conducted a review of the social network paradigm and summarized four canonical types of study based on network outcomes: structural

DOI: 10.1201/9781003289845-10

social capital, social access to resources, contagion, and environmental shaping. However, although the volume of social network research in management has increased radically in recent years, SNA is not common in studies about workplace environments and their management. So far, the research that most often applies SNA methods is the ones exploring the spatial proximity effect on building occupants' social interactions (Kabo, 2017, 2018; Zhou & Hua, 2020). The goal of this chapter is to summarize the methodological approaches using SNA methods, taking spatial proximity studies as an example, and propose directions for future research.

10.2   Argument

With the recent trends of new workplace practices such as remote working coworking, an essential question that needs to be more explored is whether and how workplace spatial factors affect human social interactions and networking. However, how workplace environment affects communication and social interactions in organizations is a challenging topic as it requires an understanding of the socialization process, relations, and outcomes that could not usually be studied at the individual level. This type of study goes beyond the approach of user experience studies which are commonly based on individual users' perceptions. In order to understand social dynamics in a certain group, the appropriate method that investigates social interactions and relations ought to be applied. The central argument of this chapter is that SNA can contribute to workplace studies by adding a new perspective and analytical methods to study employees' social behavior in the workplace. SNA in this case offers a new perspective and a set of methods to collect and look into relational data, which is distinct from data or attributions focused mainly on the individual's characteristics.

10.3   Examples of application

By operationalizing spatial factors and other social relations as relational data, a group of studies have explored social interactions and group dynamics in the workplace through SNA. Taking spatial proximity as an example, proximity refers to the spatial distance between occupants and it was found to affect unplanned encounters in the workplace (Hiller et al., 1987; Peponis et al., 2007; Shpuza & Peponis, 2008). A plethora of studies has found spatial proximity affects the formation of social relationships in the workplace and other situations (Allen, 2007; Kabo, 2017; Kleinbaum et al., 2013; Mok & Wellman, 2007; Yuan & Gay, 2006). This chapter showcases examples of how to quantify spatial factors and social factors in relational data and the general process of conducting an SNA study.

10.3.1   Spatial proximity as relational data

The spatial distance between any two individuals, as a kind of relational data, is a common measure used to understand how workspace affects employees' encounters and social interactions in recent spatial proximity studies. Kabo (2018) measured spatial distance in a concept named integration (the average spatial distance of an individual's office space relative to all other spaces) and found that, for individuals, there is a significant correlation between spatial distance and social interaction. Other types of spatial proximity measures, including dyadic physical distance and functional zone overlap, were also significantly and positively correlated with the chance of collaboration among the researchers in a research facility (Kabo et al., 2014, 2015). In an office setting, the spatial distance between any two individuals was found to be negatively and significantly related to potential face-to-face encounters. At the same time, other demographic homophily variables are non-significant (Kabo, 2017). In summary (see Table 10.1), testing the correlation between spatial distance and communication/interaction frequency at a dyadic level shows a new approach to empirically studying the spatial proximity effect on interaction.

Table 10.1 Variables and Data Analysis Methods

References	Independent Variables	Dependent Variables	Data Analysis Methods
Zhou and Hua, (2020)	Perceptual co-presence; Co-presence based on room access history	Friendship and advice network relations	Multiple-Regression Quadratic Assignment Procedure (MQAP) Analysis
Kabo, (2018)	Spatial layout network (integration/distance), interaction network: degree and betweenness centrality	Network collective intelligence; Prestige or status outcome	Heckman sample selection regression models; Multilevel mixed-effect passion model
Kabo, (2017)	Homophily (tenure, gender, education); structure (staff/ manager, group affiliation); proximity (office distance)	Potential encounters (co-presence in different distances)	Quadratic Assignment Procedure (QAP) Analysis
Kabo et al. (2015)	Path overlap; Physical distance between office	Collaboration index (IRB, Research proposal applications)	Zero-inflated negative binomial regression
Wineman et al. (2014)	Space layout (MetChoice; MeanDist); Degree centrality, Betweenness centrality	Innovation involvement; publications	Logistic Regression; Negative Binomial
Wineman et al. (2009)	Department affiliation; Spatial distance (based on depth map)	Co-authorship	Logit Regression
Peponis et al. (2007)	Density of interaction (interact network size & frequency)	Time spent on Productivity (one of the five categories)	Correlation analysis; Space syntax
Rashid et al. (2006)	Space layout variables: Integration, connectivity	Interaction, co-presence, movement	Correlation analysis; Space syntax

In general, spatial distance is an effective proxy for interaction in traditional workspace and arrangement, meaning when people work at an assigned desk and go to the same office every day, spending the entire workday there. However, given new and more flexible ways of working, spatial distance measure shows limitations. Suppose a worker takes a significant amount of time working remotely. In that case, the distance between the assigned office/workstation locations may not accurately reflect the chance for this individual to interact with the neighboring workers. This measure is also challenging for quantifying proximity between workers working in a shared space without assigned desks (e.g., hoteling or hot-desking policies). Because humans have the agency to move, without the help of advanced indoor location positioning systems, measuring the exact distance between occupants remains a challenge.

Instead of measuring the physical distance between the workstations, co-presence is a temporal relational concept that measures how much time any two individuals are in the same space. Hillier et al. (1987) argued that spatial layout in itself generates a field of probabilistic co-presence and encounter. The co-presence pattern has both a describable pattern and a known cause, which the author called the "virtual community" (Hillier et al., 1987). Compared to spatial distance between offices or workstations, co-presence provides another way to represent spatial proximity within a flexible working environment. It is more suitable for studying the chance of social encounters in workplace settings such as hot-desking.

Zhou and Hua (2020) explored whether the use of a shared study space played a role in shaping graduate students' social networks by studying how the co-presence in a shared space was related to the structure of friendship and advice networks. Applying SNA allows the authors to explore the detailed relational data in the whole student group, and conduct a correlational study to understand how space affects social connections in a setting that is not occupied in a fixed schedule. The authors also argued that the increasing data availability in facilities management heralds exciting possibilities for empirically modeling co-presence in small spaces. By calculating "space occupancy time overlap between the occupants," the longitudinal behavioral data obtained from various sensor records could be used to study the proximity effect in the workplace in a new spatial-temporal approach.

10.3.2 Social characteristics as relational data

Known as "birds of a feather always flocked together," homophily is the principle that contact between similar people occurs at a higher rate than among dissimilar people (McPherson et al., 2001). Social homophily as a network variable describes whether any two persons in a network share the same social characteristics such as race, gender, age, culture, group affiliation, etc. Homophily studies have dealt widely with various kinds of social relations such as marriage, friendship, advice networks, managers' instrumental networks in organizations, business relationships, job referrals, and so on (Kilduff & Tsai, 2003, p. 52).

In these studies, social homophily network is commonly constructed as a matrix with numbers of "1" or "0" to represent whether any two actors shared the same characteristics such as gender, ethnicity, or have attended the same event, etc. The goal is to use a matrix to show the similarity between any two actors regarding a specific characteristic. For data analysis, previous studies have used Quadratic Assignment Procedure (QAP) and Multiple-Regression Quadratic Assignment Procedure (MQAP) analysis to explore the correlation between matrices to understand factors (such as proximity, social homophily, etc.,) that affect network formation. In a research institution, Kegen (2013) found that there is no significant gender homophily effect but rather proximity effect on research cooperation, research support, and social acquaintance. It helps to differentiate the proximity from social homophily effect on interactions and communication patterns. In a study about the homophily of network ties in distributed teams, Yuan and Gay (2006) found that both homophily in group assignments and in location had significant impacts on the development of network ties. SNA allows a researcher to study social characteristics in a form of matrices that can be constructed through individuals' similarities or differences in social characteristics.

10.3.3 Summary of process

10.3.3.1 Identify network variables

Each research design for SNA starts with defining what kind of data and for what purpose will be gathered. The first step is to understand the main goal of the research as well as specify the independent variable or dependent variables that describe relations between actors. Besides the general research design process, researchers and practitioners need to understand relational data are unique, and not every research needs relational data. Research questions related to experiential factors such as work environment satisfaction, commonly based on an individual's experience, could be studied without including relational data. However, topics related to the number of friends, who collaborates with whom (e.g., co-authorship), and communication patterns can benefit significantly by understanding the specific network of friendship, advice, communication, etc. Besides, it is also important to understand that network ties sometimes have direction (e.g., who sends messages to whom) and value (e.g., the strength of friendship tie). While designing the survey questions, it is also important to include questions regarding the actors' and ties' specific characteristics besides questions about relations.

10.3.3.2 Selecting a proper sample

Selecting a proper sample is especially important when the data is collected using surveys, questionnaires, or observations. So far, most SNA studies in workplace research use a purposive sampling strategy and try to obtain every actor's response to construct a whole network. The researcher needs to understand that collecting data from every actor is important, and losing data from important ones might significantly affect the study's final results. In such a situation, researchers need to make sure to have access to collect information from every actor in the network, and be aware of the limited capacity and time when it comes to collecting data. The sampling procedure and targeted sample size should be defined before the data collection will be performed.

10.3.3.3 Data collection

SNA research has identified three types of data also called units of analysis, which should be and are investigated: relations, ties (Garton et al., 1997), and actors. Many methods of obtaining network data such as questionnaires, interviews, observations, and artifacts exist (Garton et al., 1997). Wasserman and Faust's (1994) book provides comprehensive guidance regarding data collection tools and methods for SNA.

10.3.3.4 Data preparation

Network data analysis puts emphasis not on the individuals themselves but on the relationships among people. In contrast to data about individual characteristics, network data normally consists of a rectangular array of measurements. Collected data has to be represented in a way that facilitates the application of SNA methods. A widespread representation is graph or matrix, which can be done manually by extracting information from surveys, interviews, and observations or automatically using data-mining techniques for data cleaning. The goal of data preparation is to represent the collected data in the form of a network.

10.3.3.5 Applying appropriate analytical methods

Taking spatial proximity studies as an example, three types of data analysis have been explored in previous studies according to different research questions by calculating (1) the number of actors ties, (2) the correlation between network matrices, and (3) network graphs. For studies exploring how spatial factors affect individuals' social connectivity in a network, the number of ties (in networks such as friendship, advice, etc.) has been used to describe how individuals are connected to others (Tagliaro et al., 2022). Correlation analysis and regression analysis could be conducted to explore how spatial factors affect an individual's social connectivity. In some other studies, network variable is simplified as "1" or "0" to represent whether there is a relationship or not, such as co-authorship (Wineman et al., 2009, 2014). The logit regression is a common analytic method in this case.

For research that explores correlations between networks/matrices, such as how spatial proximity network is correlated with friendship network, methods like QAP and MQAP are commonly used (examples shown in Table 10.2 and Table 10.3) and they offer an opportunity to compare proximity effect with other social homophily effects such as gender and ethnicity (Kabo, 2017; Zhou & Hua, 2020). QAP

Table 10.2 Example of QAP analysis that calculates the correlation between different matrices by using quadratic assignment procedures to develop standard errors to test for the significance of association (Zhou & Hua, 2020)

	QAP correlation between networks									
	F(s)	*F(w)*	*Advice*	*Co (sur)*	*Card (s)*	*Social*	*Class*	*S-gen*	*S-nat*	*S-coh*
Friendship (strong)	1.00									
Friendship (weak)	−0.16**	1.00								
Advice	0.52**	0.17**	1.00							
Copresence (survey)	0.32**	0.20**	0.47**	1.00						
Card access (same-day)	0.23**	0.18*	0.47**	0.59**	1.00					
Social media	0.64**	0.17**	0.62**	0.40**	0.38**	1.00				
Class collaboration	0.40**	0.12*	0.37**	0.28**	0.15*	0.46**	1.00			
Same-gender	−0.04	−0.13*	–	−0.06	−0.02	0.02	–	1.00		
Same-nationality	0.13*	0.13**	0.25**	0.13**	0.22**	0.22**	0.06	−0.03	1.00	
Same-cohort	0.39**	0.03	0.23**	0.13**	−0.03	0.27**	0.15**	0.03	–	1.00

Note: ** $p < .01$; * $p < .05$.

Table 10.3 Example of MQAP analysis that shows when multiple matrices were entered into the regression to see each matrix's effect while controlling others. For example, the Model 2 (M2) for predicting Friendship (strong) relations suggests that the standardized regression coefficient for the impact of social media on strong friendship was 0.46 ($p<0.01$), which was significantly higher than Copresence (survey) 0.07 ($p>0.05$) (Zhou & Hua, 2020)

MQAP regression results

	Friendship (strong)				Friendship (weak)				Advice			
	M1	M2	M3	M4	M1	M2	M3	M4	M1	M2	M3	M4
Intercept	0.10**	-0.02**	-0.02**	-0.08**	0.21**	0.19**	0.19**	0.17**	0.14**	0.03**	0.03**	-0.02**
Copresence (survey)	0.30**	0.07	0.06	0.05	0.19**	0.15*	0.15*	0.15*	0.51**	0.29**	0.28**	0.27**
Social media		0.48**	0.43**	0.39**		0.08*	0.07	0.06		0.46**	0.43**	0.39**
Class collaboration			0.13**	0.12**			0.03	0.04			0.09	0.10
Same-nationality				0.01				0.06*				0.09**
Same-cohort				0.15**				-0.01				0.05
R-square	0.10	0.41	0.42	0.47	0.04	0.05	0.05	0.06	0.22	0.44	0.45	0.46
Observations	702	702	702	702	702	702	702	702	702	702	702	702

Note: ** $p < .01$; * $p < .05$.

regression has a unique data structure in which each matrix of relations represents a variable, and analogous cells across the set of all matrices constitute a case (Krack-hardt, 1987). In both QAP and MQAP analysis, the unit of analysis is a dyad, a pair of individuals who may or may not have some sort of relation connecting them to one another. Once a dataset is assembled and an OLS regression is carried out, the resulting coefficients indicate the direction of effect of independent variables upon the dependent variable.

The network graph visualization shows undirected and directed graph structures and is another common way for network visualization. This type of visualization illuminates relationships between actors and entities, and it can be generated through software such as Gephi, UCINET, and coding software such as igraph and RStudio. Zhou et al. (2021a) used a network graph to describe the groups' spatial adjacency preference in a network format during an organization's space planning (shown in Figure 10.1). They argue that such a network graph is good at showing the patterns of the group's location in a network and its relationships with the others. This information can be helpful to inform spatial choices regarding interior space layout and people's distribution within the building.

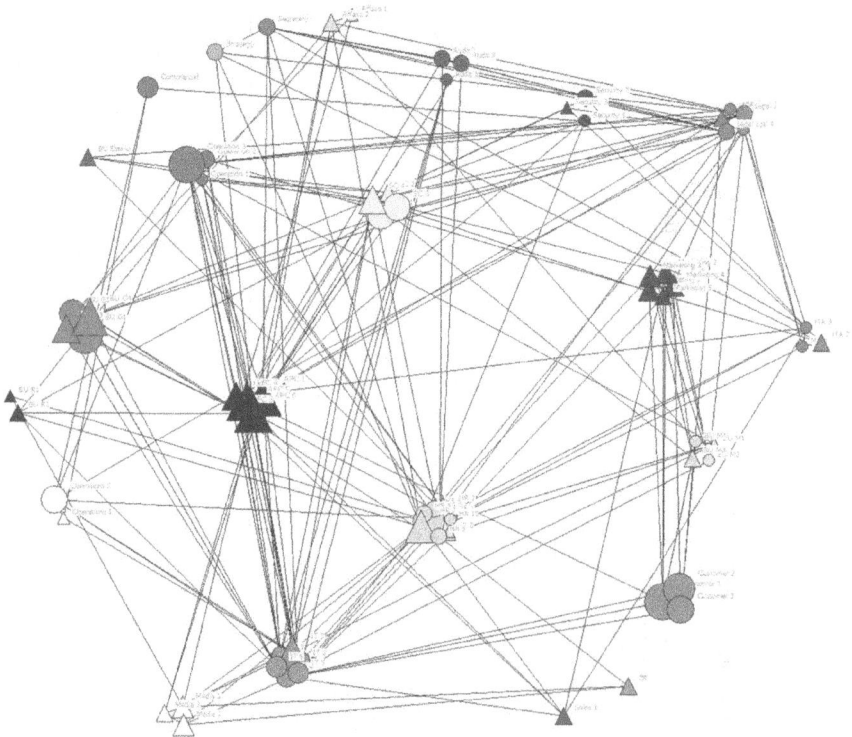

Figure 10.1 Network diagram of workgroup's spatial adjacency preference (Zhou et al., 2021a)

10.4 Implication

10.4.1 A SNA approach to study how space affects communication in workplace

The network approach to studying proximity effects provides opportunities to combine spatial and social network analyses. A five-step approach to conduct a workplace study based on SNA method is proposed by the author (shown in Figure 10.2). As discussed in Section 3.1, both spatial distance and co-presence relationships at a dyadic level can help researchers explore communication patterns in the workplace more detailedly.

For dependent variables, depending on the research questions, network variables such as degree centrality and closeness centrality could be used to describe an individual or group's connection to others in the communication/interaction network. Other data sources such as the building occupancy and email exchange records which reflect an organization's communication patterns become feasible to obtain nowadays. They suggest a promising new category of user behavior information to better understand how space aligns with the actual workflow and collective behavior.

For independent variables, co-presence has been explored as a new type of variable for quantifying proximity in a flexible or shared work environment. With the development in sensor technologies, more sensor-based tools are now available for measuring building occupants' presence and movement in an indoor environment.

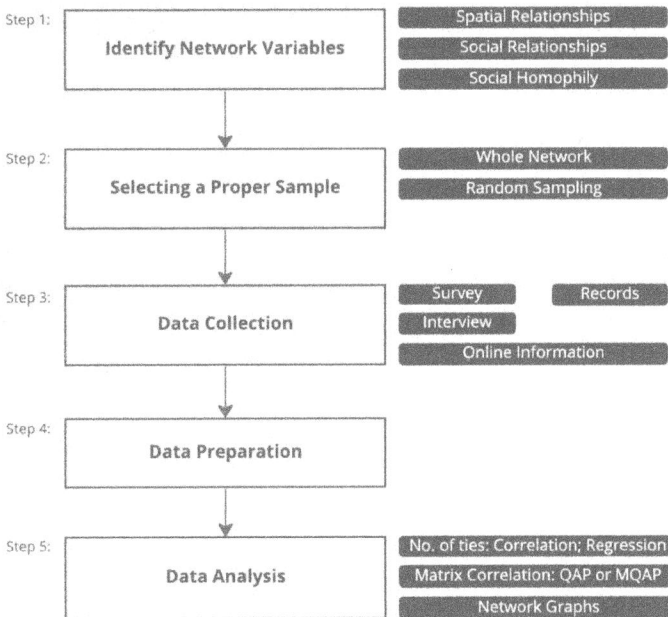

Step 1: **Identify Network Variables**
- Spatial Relationships
- Social Relationships
- Social Homophily

Step 2: **Selecting a Proper Sample**
- Whole Network
- Random Sampling

Step 3: **Data Collection**
- Survey
- Interview
- Records
- Online Information

Step 4: **Data Preparation**

Step 5: **Data Analysis**
- No. of ties: Correlation; Regression
- Matrix Correlation: QAP or MQAP
- Network Graphs

Figure 10.2 Summary of conducting a SNA study

Multiple review articles have been published recently regarding different categories of occupancy detection technology used in previous studies and discuss their pros and cons (Azimi & O'Brien, 2022; Zhou et al., 2021a).

10.4.2 A SNA approach to study spatial adjacency preferences for space planning

During office space planning, there is a common strategy to construct a spatial adjacency diagram, or a "bubble" diagram, which indicates the desired circulation connections between spaces to achieve optimal work efficiency or other goals, such as facilitating communication and collaborations (Cherry & Petronis, 2007). This diagram synthesizes the required or preferred spatial proximity information obtained from the end users and provides recommendations for the spatial layout design. Questionnaires and interview instruments were developed to collect spatial adjacency preferences in previous case studies of large organizations' office planning (Preiser, 1993).

SNA software such as UCINET and Gephi are appropriate tools for processing and visualizing data indicating communication patterns and spatial preferences during space planning for large organizations. Identifying the central groups in a collaboration and communication network can help detect the workgroups that are suitable for being located at the center of a floor plan or close to core areas, which would increase the accessibility for the other groups (Zhou et al., 2021b). Network graphs also show exciting possibilities for space planners based on a deeper understanding of the organization's communication pattern. They are good at showing the patterns of the group's relationships with the others; they might be helpful in informing interior space layout and people's distribution within the building.

10.5   Conclusions

SNA allows researchers to conceptualize concepts such as spatial proximity, social interaction, and social homophily as relations that could be studied through networks. It provides the opportunity to explore the confounding relationships between workspace and the other social characteristics variables and allows the exploration of new relational measures such as co-presence in time. The SNA approach to study spatial proximity effect in the workplace contributes to answering the question raised by Gans (2002, p. 133) regarding space and human behavior: "do both natural and social space have casual power, creating social effects, and if so when, how, and why?"

Future studies can explore applying SNA methods to further study how space affects employees' communication and interaction patterns in workplace. Several limitations and challenges are identified when applying SNA:

1 *Network changes as time passes by.*
    It has to be acknowledged that social network is not static and any social network analysis at a specific time only shows evidence describing the social relations to

the particular moment. To study how the spatial proximity effect affects social interaction through SNA, future studies need to consider the evolving character of social networks. Multiple rounds of network data collection, especially before and after the change in spatial factors, will be beneficial for studies showing evidence of social network changes.

2 *Considering virtual communication.*

With the broad adoption of remote working, social interactions in the workplace have become growingly dynamic. Studying workplace spatial factors in such a flexible context is a timely topic and requires more diverse methods to explore workplace design and space planning. Studies have suggested that interactions in physical space and virtual communication channels are not exclusive but interrelated (Zhou & Hua, 2020). It would be interesting to explore the role of spatial factors' role in social interactions and how they interact with other online communication channels.

10.6   Further readings

There are some excellent handbooks written by Wasserman and Faust (1994) and Domínguez and Hollstein (2014) introducing various SNA methods and describing all the steps involved in data collection and statistical analysis:

Wasserman, S., & Faust, K. (1994). *Social network analysis: Methods and applications.* Cambridge University Press.
Domínguez, S., & Hollstein, B. (Eds.). (2014). *Mixed methods social networks research: Design and applications*. Cambridge University Press.

In addition to these handbooks, Nohria et al. (1992) and Kilduff and Tsai's (2003) books introduced networks research in organizational studies which covers information about both theories and methods:

Nohria, N., Eccles, R. G., & Press, H. B. (Eds.). (1992). *Networks and organizations: Structure, form, and action.* Harvard Business School Press.
Kilduff, M., & Tsai, W. (2003). *Social networks and organizations*. Sage.

References

Allen, T. J. (2007). Architecture and communication among product development engineers. *California Management Review, 49*(2), 23–41.
Azimi, S., & O'Brien, W. (2022). Fit-for-purpose: Measuring occupancy to support commercial building operations: A review. *Building and Environment, 212*, 108767. https://doi.org/10.1016/j.buildenv.2022.108767
Borgatti, S. P., & Foster, P. C. (2003). The network paradigm in organizational research: A review and typology. *Journal of Management, 29*(6), 991–1013.
Breiger, R. L. (2004). The analysis of social networks. In M. Hardy, & A. Bryman (Eds.), *Handbook of data analysis* (pp. 505–526). Sage.

Cherry, E., & Petronis, J. (2007), *Architectural programming*, World Building Design Guide, Institute of Building Sciences, available at: www.wbdg.org/design-disciplines/architectural-programming (accessed 4 March 2023).

Gans, H. J. (2002). The sociology of space: A use–centered view. *City & Community*, *1*(4), 329–339.

Garton, L., Haythornthwaite, C., & Wellman, B. (1997). Studying online social networks. *Journal of Computer-Mediated Communication*, *3*(1), JCMC313.

Granovetter, M. (1983). The strength of weak ties: A network theory revisited. *Sociological Theory*, *1*, 201–233.

Hillier, B., Burdett, R., Peponis, J., & Penn, A. (1987). Creating life: Or, does architecture determine anything? *Architecture et Comportement/Architecture and Behaviour*, *3*(3), 233–250.

Kabo, F. (2017). A model of potential encounters in the workplace: The relationships of homophily, spatial distance, organizational structure, and perceived networks. *Environment and Behavior*, *49*(6), 638–662. https://doi.org/10.1177/0013916516658501

Kabo, F. (2018). The architecture of network collective intelligence: Correlations between social network structure, spatial layout and prestige outcomes in an office. *Philosophical Transactions of the Royal Society B: Biological Sciences*, *373*(1753), 20170238.

Kabo, F. W., Cotton-Nessler, N., Hwang, Y., Levenstein, M. C., & Owen-Smith, J. (2014). Proximity effects on the dynamics and outcomes of scientific collaborations. *Research Policy*, *43*(9), 1469–1485.

Kabo, F., Hwang, Y., Levenstein, M., & Owen-Smith, J. (2015). Shared paths to the lab: A sociospatial network analysis of collaboration. *Environment and Behavior*, *47*(1), 57–84.

Kegen, N. V. (2013). Science networks in cutting-edge research institutions: Gender homophily and embeddedness in formal and informal networks. *Procedia - Social and Behavioral Sciences*, *79*, 62–81. https://doi.org/10.1016/j.sbspro.2013.05.057

Kilduff, M., & Tsai, W. (2003). *Social networks and organizations*. Sage.

Kleinbaum, A. M., Stuart, T. E., & Tushman, M. L. (2013). Discretion within constraint: Homophily and structure in a formal organization. *Organization Science*, *24*(5), 1316–1336. https://doi.org/10.1287/orsc.1120.0804

Krackhardt, D. (1987). QAP partialling as a test of spuriousness. *Social Networks*, *9*(2), 171–186.

McPherson, M., Smith-Lovin, L., & Cook, J. M. (2001). Birds of a feather: Homophily in social networks. *Annual Review of Sociology*, *27*(1), 415–444.

Mok, D., & Wellman, B. (2007). Did distance matter before the Internet?: Interpersonal contact and support in the 1970s. *Social Networks*, *29*(3), 430–461.

Peponis, J., Bafna, S., Bajaj, R., Bromberg, J., Congdon, C., Rashid, M., Warmels, S., Zhang, Y., & Zimring, C. (2007). Designing space to support knowledge work. *Environment and Behavior*, *39*(6), 815–840. https://doi.org/10.1177/0013916506297216

Pool, I., & Kochen, M. (1978). Contacts and influence. *Social Networks*, *1*(1), 5–51.

Preiser, W.F.E. (1993). *Professional practice in facility programming*. Van Nostrand Reinhold.

Rashid, M., Kampschroer, K., Wineman, J., & Zimring, C. (2006). Spatial layout and face-to-face interaction in offices—a study of the mechanisms of spatial effects on face-to-face interaction. *Environment and Planning B: Planning and Design*, *33*(6), 825–844.

Scott, J., & Carrington, P. J. (2011). *The SAGE handbook of social network analysis*. Sage.

Shpuza, E., & Peponis, J. (2008). The effect of floorplate shape upon office layout integration. *Environment & Planning B, Planning & Design*, *35*(2), 318–336.

Tagliaro, C., Zhou, Y., & Hua, Y. (2022). Work activity pattern and collaboration network: New drivers for workplace space planning and design. *Journal of Interior Design, 47*(3), 29–46.

Wasserman, S., & Faust, K. (1994). *Social network analysis: Methods and applications.* Cambridge University Press.

Wineman, J., Hwang, Y., Kabo, F., Owen-Smith, J., & Davis, G. F. (2014). Spatial layout, social structure, and innovation in organizations. *Environment and Planning B: Planning and Design, 41*(6), 1100–1112.

Wineman, J. D., Kabo, F. W., & Davis, G. F. (2009). Spatial and social networks in organizational innovation. *Environment and behavior, 41*(3), 427–442.

Yuan, Y. C., & Gay, G. (2006). Homophily of network ties and bonding and bridging social capital in computer-mediated distributed teams. *Journal of Computer-Mediated Communication, 11*(4), 1062–1084.

Zhou, Y., & Hua, Y. (2020). The role of shared study space in shaping graduate students' social networks. *Journal of Facilities Management, 19*(1), 92–110. https://doi.org/10.1108/JFM-08-2020-0060

Zhou, Y., Hua, Y., & Liu, J. (2021a). Study workplace space occupancy: A review of measures and technologies. *Journal of Facilities Management, 20*(3), 350–368. https://doi.org/10.1108/JFM-01-2021-0013

Zhou, Y., Tagliaro, C., & Hua, Y. (2021b). Networked "bubbles": Study workgroups' spatial adjacency preference using social network analysis methods. *Journal of Corporate Real Estate, 23*(2), 87–105. https://doi.org/10.1108/JCRE-06-2020-0024

11 Surveys

Gathering data for workplace post-occupancy evaluation

Ying Hua

Cornell University, USA

11.1 Background

Post-Occupancy Evaluation (POE) is a process of evaluating buildings after they have been built and occupied for some time, with a focus on building occupants and their needs. It was defined as a systematic diagnosis procedure to evaluate critical aspects of building performance to form a sound basis for improvement or better future design (Cooper, 2001; Preiser, 1995; Preiser et al., 1988). Though POE is most commonly used during the building's operation stage as a feedback process, it can also be used during the pre-project as a feed-forward method or as a benchmarking method for the evaluation of buildings and facilities to maintain effectiveness (Cooper, 1985). Hadjri and Crozier (2009) further stated that the goal of POE is to facilitate the accumulation of information that can be subsequently utilized to improve the procurement of buildings to the benefit of all stakeholders involved.

Some of the methods used in POE are objective measurements of the physical building environment, like various Indoor Environment Quality (IEQ) *in-situ* measurements; some are subjective measurements like occupant surveys, interviews, walkthroughs, etc.; and surveys are the most widely used method to capture occupant experience and feedback (Dykes & Baird, 2013; Li et al., 2018). A survey is a research method that studies the characteristics of a given population through respondents' self-report data collection. It can be generally explained as "questioning individuals on a topic or topics and then describing their responses" (Jackson, 2011, p. 17). It has long been a central strategy in social science research (Punch, 2003). Since it would be challenging to reach every subject in the entire population, a survey is often conducted with a sample of that population. It requires a careful design of sampling frames and recruitment strategies for the accuracy of the survey data and the robustness of the survey results to represent the population under study (Chaudhuri & Stenger, 2005; Kalton, 2021).

Surveys are a major source of quantitative data on occupants in POE studies and are usually carried out in the form of questionnaires distributed to the targeted sample. Qualitative data could also be collected via open-ended questions in surveys. In the context of the workplace, POE survey questions are frequently used to examine occupant satisfaction with IEQ, including thermal comfort, satisfaction with indoor air quality, visual comfort, acoustic comfort, etc. (Galatioto

DOI: 10.1201/9781003289845-11

et al. 2013). Spatial quality is among the key IEQ aspects of the workplace, but is less extensively studied, whereas occupant perception of spatial quality and the perceived support from the environment for work experience, work performance, and other outcomes are central to workplace studies. The application of surveys in workplace research and management reflects the person-environment (P-E) fit theory (see Caplan, 1987), which makes the probing of occupant experience and the development of survey instruments and processes an intriguing research topic – see the Further Reading section below to expand on the subject of P-E fit with the workplace (Armitage & Amar, 2021).

Survey design has several important reliability and validity issues to address, including but not limited to test-retest reliability, parallel (or alternate) forms reliability, and internal consistency (Iarossi, 2006; Litwin, 1995). Besides, the length of the survey, i.e., time and effort needed for survey respondents to complete a survey, order of questions, and choice of the rating scale, all need to be carefully thought through by researchers to the goals of the study, population studied, and more sector context information, besides survey design principles, to ensure that the quality of the data collected is reliable; therefore the results are meaningful. In terms of media, looking at current practice, workplace surveys can be paper-based or web-based, each with obvious strengths and weaknesses (Wright, 2005). There is a tendency for more surveys being conducted online and more robust online survey tools and platforms being developed – this is one of the major new developments of the survey as a classic research method with a long history. Both survey design and survey media issues will be elaborated more in the sections below.

11.2 Argument

Researchers are using survey method in workplace POE studies to collect a wide variety of data from workplace occupants (Bunn & Marjanovic-Halburd, 2017; Candido et al., 2016; Gocer et al., 2018; Hua, 2010; Hua et al., 2010; Hua et al., 2011; Hua et al., 2014; Kim & De Dear, 2012; Lee et al., 2020; Xu et al., 2016). These data sets include:

- Work environment satisfaction indicates satisfaction with general and/or specific aspects of the work environment. The term environment here takes a broad definition. On the one hand, it includes the physical work environment such as location, layout, spatial features, furniture, and various aspects of IEQ. On the other hand, the environment here can also indicate non-physical aspects of work, including social, managerial, operational, and technical environment at work.
- General perception indicates general satisfaction and other perceived outcomes including health, well-being, and productivity (could be individual level and group level).
- Work experience indicates satisfaction with key work processes, information flow, interface with colleagues, etc.

- Work behaviour and perceived patterns report respondents' behaviour such as choice of settings or seat selection, nature of tasks, etc.
- Personal traits range from respondents' demographic information to personality information.

Based on data collected by surveys and connecting different data sets outlined above, studies used several ways to examine the relationship between workplace and their occupants' perception and experience in studies of various types – descriptive, exploratory, explanatory, or evaluation research – to answer different kinds of research questions and test different types of hypotheses. Such use of data collected by surveys goes beyond using descriptive statistics to paint a picture of what is going on in a workplace in terms of occupant satisfaction. Still, it enables the use of sophisticated and systematic statistical tools to probe much more profound and to address workplace phenomena at a theoretical level:

- Mapping of environmental satisfaction and occupants' activities in the workplace to understand behaviour's spatial patterns.
- Testing the difference between perceptions, experience, and outcomes measurements by different occupant subgroups or at other times (e.g., in a before-and-after study) to understand the impact of design or management interventions.
- Building prediction models by regression analysis to develop theories.
- In the new hybrid work mode that is demonstrating a noticeably higher proportion of time working from locations that are not company office spaces than before – understand the fundamental changes in occupants' behaviour and expectations, generate insights about the mechanism to inform the practice of transferring the best practice to other work locations to continue support occupants and businesses.

As a tool that relies on respondents' commitment of time and interest for the volume and quality of data obtained, surveys have some intrinsic challenges, especially when work itself and means of collecting data in the workplace become more virtual instead of face-to-face. The length of the survey and the level of difficulty of questions are among the factors that directly influence the survey response rate. On the one hand, it requires us to expand our knowledge of methods of occupant data collection and pick the right combination of tools, which speaks of the value of this book on workplace study and a wide array of data collection methods. Compared to before, we have more ways to document occupant behaviour beyond relying on self-report. More elaboration will be made on this topic later in this chapter.

On the other hand, although people seem to have less tolerance for survey tasks in the digital age, the move of the survey from paper-based to web-based does give us more tools. Technically speaking, some online survey platforms allow the presentation of the survey questions to be tailored according to participants' responses throughout the survey. Such a function eliminates the time spent on irrelevant content on the survey and reduces the amount of time spent or the length of the survey, hence supporting the survey completion rate. This function also creates an opportunity to probe into more details, including reasons for dissatisfaction, i.e., when an answer to the main question is below neutral, or gather more information about features that are particularly meaningful and favoured by survey respondents when an answer to a certain question is above neutral level, without the need to add probing questions for all participants.

There are also new challenges for POE surveys that have emerged and enlarged along with the evolution of office work toward more digital and hybrid. A major one is a conflict between the relatively static nature of the survey tool and the increasingly fluid and mobile nature of work. It does not dismiss the value and importance of survey as a workplace study tool, but calls for new techniques and new combinations with other methods for robust workplace studies. The discussion on this topic will continue in the following sections in this chapter, as it reflects the most significant change in work and workplace, the lift of spatial-temporal constraints by technology and management.

11.3   Example of application/use

An example of a *workplace collaborative environment survey* will be presented in this section, illustrating a considerable emphasis on survey studies of the workplace – focusing on workplace support for collaboration and collaboration experience, as workplace's facilitation for co-creation and social interaction is a highlighted function in the hybrid work era.

This is a survey created for studying collaborative work environments. The initial idea and effort to develop this survey started around 2004 as a key instrument for doctoral thesis work on this topic. The survey has been tested and strengthened over a long time of application in multiple research projects of different goals, including performance diagnosis in workplace POE, empirical study for theoretical model building, and evidence-based workplace consulting work.

The first version of the survey (Hua, 2010) consisted of five parts:

i Occupant satisfaction with the work environment's support for collaboration (18 items).
ii Spatial preference for interaction and cooperation.
iii Occupant satisfaction with collaboration-related work experience (12 items).
iv Occupant perceived collaboration effectiveness (4 items).
v Demographic characteristics.

The 18 statements in Part I cover the amount, size, location, combination, and availability of various formal and informal collaborative spaces at work, including meeting rooms/spaces, common copier/printer areas, shared kitchen and coffee areas, circulation areas, and individual workstations, as well as the furnishings, tools, and technologies in the spaces mentioned above. Participants were also asked about their overall satisfaction with the quality of their work environment in terms of whether or not it supported their desired interactions and collaborations for work deliverables. Five-point Likert scale from "strongly disagree" (1) to "strongly agree" (5) was used.

Exploratory and confirmatory factor analyses of the 18 items revealed a two-factor model of occupant satisfaction with a collaborative work environment without cross-loading. The two factors are perceived facilitation from the work environment for collaboration and perceived distraction from the work environment because of interactive behaviors (factor loading cut-off at 0.40, Cronbach's alpha of the two underlying factors at 0.90 and 0.79).

Data from a sample of 308 participants from multiple organizations and sites (but all US federal government offices) were used in a structural equation model to look at the relationship between workplace spatial settings and workplace environment satisfaction, collaboration experience and perceived effectiveness at work. Besides data from the survey, a series of spatial data were collected for each participant's office or cubicle, i.e., its distance from various shared spaces on the workplace floor and the collaborative space amount and distribution on the floor. Based on these data, a prediction model was proposed, in which workstation-meeting space distance, workstation-kitchen distance, percentage of space on the floor dedicated to shared services, and amenities predicted the occupant perceived facilitation from the spatial work environment for collaboration (Hua et al., 2011). This occupant and spatial data set also enabled a typology study of shared spaces on 27 office floors (Hua et al., 2010).

Over time, the survey has been expanded and adjusted in three ways – expanded workplace environmental satisfaction section (total of 32 items) with a new factor loading structure, more detailed aspects of work experience, and added personality section for the possibility of more in-depth understanding of occupant responses to a workplace setting changes. Below is the new structure with factor-loading information:

1 Satisfaction with the work environment.

   • Support from the work environment for collaboration
   • Support from individual workspace for solo work
   • Distraction from shared spaces in the workplace
   • Reluctance to interact in shared space in the workplace

2 Work engagement.
3 Work experience.

   • Information flow
   • Socialization
   • Team coordination

4 Personality.

   • Extraversion
   • Conscientiousness
   • Emotional stability
   • Agreeableness
   • Openness to experience

A later application of this survey in a study which has a more practical goal of informing workplace design and strategy added another Nature of Work section seeking to get an understanding of how employees typically spend their time at work, what media they use, what percentage of time they spent on individual and collaborative work, etc. This new section brought valuable information for the analysis and diagnosis. However, it is also a relatively difficult set of questions for participants to complete, as it requires a clear memory of work activities of the day and the week retrospectively.

Besides the robustness of this survey, another feature of this series of workplace studies enabled some valuable insights and theoretical models. That is the collection of location information and spatial characteristics of participants' offices or cubicles. It allowed the link between perception data and physical spatial data, further inspiring another set of spatial satisfaction mapping studies (Hua et al., 2014). Currently, this survey continues to be used for collecting more empirical evidence on collaboration in office space to test the existing theories under changing contexts with a higher level of hybrid work mode. It is also used to inform the design of other survey instruments, such as the one used for the study of work-from-home experience surveys.

11.4 Implications: Method relevance to research and practice

Survey has been a powerful and highly frequently used tool for workplace studies, helping researchers to collect large volumes of data for quantitative analysis and managers to back their workplace strategies with strong evidence from their companies' populations. Besides the need to continue focusing on ensuring the validity and reliability of survey instruments in research projects, strategies to support response rate and quality of answers will continue to be useful. In addition, noteworthy is that the current trend of increased mobility of occupants in the workplace and the hybrid mode of work pose new and critical challenges for this method.

Survey results tend to create a snapshot of the occupant perception in the workplace, or in some cases a general level of satisfaction with different aspects of the work environment. First of all, in general, data are not easy to collect in real-time using one-time surveys. An aggregated and average level of satisfaction will have less meaning in the research context. Secondly, questions on occupants' perceptions lack the mechanism to capture the context information for participants' perceptions, such as activities and locations, social settings, etc., and will need new development to address. The example of the study described in the above section made an effort to record each survey respondent's location, enabling a set of spatial variables to be attached to each survey for further correlation studies. Thirdly, the participants' compliance and carefulness when answering survey questions with such increasingly complex work mode as the context may lead to burden or low quality and inconsistency of the survey data which would harm the validity of a study and will be very difficult to check.

One method may be of reference value to the future development of survey tools, or as an expansion of the survey tools currently used in workplace POE – Ecological

Momentary Assessment (EMA) was applied widely to understand the relationship between well-being and psychopathology and environmental influences by measuring within-person perceptions repeated times over time, for example, several times a day. This methodology is more conducive to capturing time- and spatially-varying explanatory factors and intraindividual fluctuations than traditional methods and thus may yield new insights into the modelling between environmental features and occupants' perceptions (Engelen et al, 2016, Engelen & Held, 2019).

EMA has more recently been extended to geographic EMA by using GPS to capture location at the time of EMA response (Zhang et al., 2020; Mennis et al., 2018). Participants are usually given a mobile phone with embedded GPS, and a brief EMA survey was administered three to six times per day over several days. Unlike traditional POE surveys, EMA surveys almost always include the subject's activity. Some EMA surveys also include location, social context, and posture. There are mainly two ways to connect environmental variables and survey data: Connecting by location – for large-scale environmental variables, when participants submit an EMA survey, their GPS information is also recorded and for environment variables; connecting by time – for microenvironmental variables, participants carry sensors and EMA tools which are usually mobile phones together and the time stamp of submitting the EMA survey can be used to map according to the environmental variables' metrics.

Analysis of social review sites is another new trend of questionnaire-based POE research. Dalton et al. (2013) proposed to apply social review sites such as Yelp and Tripadvisor to the POE study and suggested that highly valuable information was available from these sites and formed a new class of POE data. Despite the sparse of relevant studies, this topic deserves further exploration as the user-generated content provides another perspective of questionnaire-based POE and collects rich 'photovoice' data to help triangulate findings and improve validities.

11.5   Conclusions

Within the evolving context of the increasingly mobile and hybrid way of work, the survey is an area in workplace research methods calling for innovation to continue its legacy as a major method type for collecting valuable occupant data for building performance diagnosis and theory building. When workplace occupants started to routinely use multiple spaces in the office for different work tasks and activities, or work from different locations including offices, homes, and other third places, there could be a threat to internal validity if surveys are still used as a static snapshot of what's going on in the workplace and questions are not carefully crafted to capture and respond to this new way of work.

Secondly, with the emergence of more methods and tools to study behaviour and behaviour outcomes from interventions, and the adoption of these methods in workplace studies for which this book is a shred of strong evidence, it is a good time to suggest alternatives for certain parts in traditional workplace experience surveys that rely on self-report to capture occupant behaviour in the workplace, such as the percentage of time one spent on different tasks. Also, the possibility to access secondary data sets in organizations about their people and businesses should be looked into to take advantage of the potential. The goal is to identify a more robust and high-quality data source, without a heavy burden on participants.

Finally, more automated survey data collection, analysis, and visualization to assist the decision-making of designers and space managers are an area with both technological feasibility and great potential for business and social values. From manual information collection and processing to automated collection and processing, all the way to synthesizing survey data with other non-survey data sources to enable big data as information products for workplace management support, will be an exciting development to benefit multiple stakeholders and players.

11.6   Further Reading

Recommend reading Human-Computer Interaction (HCI) articles as a reference for real-time experience capturing for future survey development. Moreover, the following references can help expand on the topics of P-E fit with the workplace and the survey methodology in general.

Armitage, L. A., & Amar, J. H. N. (2021). Person-Environment Fit Theory. Application to the design of work environments. In R. Appel-Meulenbroek & V. Danivska (Eds.), *A Handbook of Theories on Designing Alignment Between People and the Office Environment* (pp. 14–26). Routledge. https://doi.org/10.1201/9781003128830-2

Denscombe, M. (2010). *The Good Research Guide for Small-Scale Social Research Projects* (4th ed.). Butterworth-Heinemann.

Groves, R., Flower, J. F. Jr., Mick, P. C., Couper, M. P., Lepkowski, J. M., Singer, E., Tourangeau, R. (2009). *Survey Methodology*. John Wiley & Sons.

References

Bunn, R., & Marjanovic-Halburd, L. (2017). Comfort signatures: How long-term studies of occupant satisfaction in office buildings reveal on-going performance. *Building Services Engineering Research and Technology*, *38*(6), 663–690. https://doi.org/10.1177/0143624417707066

Candido, C., Kim, J., de Dear, R., & Thomas, L. (2016). BOSSA: A multidimensional post-occupancy evaluation tool. *Building Research & Information*, *44*(2), 214–228. https://doi.org/10.1080/09613218.2015.1072298

Caplan, R. D. (1987). Person – Environment fit theory and organizations: Commensurate dimensions, time perspectives, and mechanisms. *Journal of Vocational Behaviour*, *31*, 248–267.

Chaudhuri, A., & Stenger, H. (2005). *Survey sampling theory and methods* (2nd ed.). CRC Press. https://doi.org/10.1201/9781420028638

Cooper, I. (1985). Teachers' assessments of primary school buildings: The role of the physical environment in education. *British Educational Research Journal*, *11*(3), 253–269. https://doi.org/10.1080/0141192850110306

Cooper, I. (2001). Post-occupancy evaluation - where are you? *Building Research & Information*, *29*(2), 158–163. https://doi.org/10.1080/09613210010016820

Dalton, R. C., Kuliga, S. F., & Hölscher, C. (2013). POE 2.0: Exploring the potential of social media for capturing unsolicited post-occupancy evaluations. *Intelligent Buildings International, 5*(3), 162–180. https://doi.org/10.1080/17508975.2013.800813

Dykes, C., & Baird, G. (2013). A review of questionnaire-based methods used for assessing and benchmarking indoor environmental quality. *Intelligent Buildings International, 5*(3), 135–149. https://doi.org/10.3390/su132414067

Engelen, L., Chau, J. Y., Burks-Young, S., & Bauman, A. (2016). Application of ecological momentary assessment in workplace health evaluation. *Health Promotion Journal of Australia, 27*(3), 259–263. https://doi:10.1071/HE16043

Engelen, L., & Held, F. (2019). Understanding the office: Using ecological momentary assessment to measure activities, posture, social interactions, mood, and work performance at the workplace. *Buildings, 9*(2), 54. https://doi.org/10.3390/buildings9020054

Galatioto, A., Leone, G., Milone, D., Pitruzzella, S., & Franzitta, V. (2013). Indoor environmental quality survey: A brief comparison between different post occupancy evaluation methods. *Advanced Materials Research, 864–867*, 1148–1152. http://doi:10.4028/www.scientific.net/AMR.864-867.1148

Göçer, Ö, Göçer, K., Ergöz Karahan, E., & İlhan Oygür, I. (2018). Exploring mobility & workplace choice in a flexible office through post-occupancy evaluation. *Ergonomics, 61*(2), 226–242. https://doi.org/10.1080/00140139.2017.1349937

Hadjri, K., & Crozier, C. (2009). Post-occupancy evaluation: Purpose, benefits and barriers. *Facilities, 27*(1/2), 21–33. https://doi.org/10.1108/02632770910923063

Hua, Y. (2010). A model of workplace environment satisfaction, collaboration experience, and perceived collaboration effectiveness: A survey instrument. *International Journal of Facility Management, 1*(2), 1–21.

Hua, Y., Göçer, Ö., & Göçer, K. (2014). Spatial mapping of occupant satisfaction and indoor environment quality in a LEED platinum campus building. *Building and Environment, 79*(1), 124–137. https://doi.org/10.1016/j.buildenv.2014.04.029

Hua, Y., Loftness, V., Heerwagen, J. H., & Powell, K. M. (2011). Relationship between workplace spatial settings and occupant-perceived support for collaboration. *Environment and Behavior, 43*(6), 807–826. https://doi.org/10.1177/0013916510364465

Hua, Y., Loftness, V., Kraut, R., & Powell, K. M. (2010). Workplace collaborative space layout typology and occupant perception of collaboration environment. *Environment and Planning B: Planning and Design, 37*(3), 429–448. https://doi.org/10.1068/b35011

Iarossi, G. (2006). *The Power of Survey Design. A User's Guide for Managing Surveys, Interpreting Results, and Influencing Respondents* [35034]. The World Bank. http://documents.worldbank.org/curated/en/726001468331753353/The-power-of-survey-design-a-users-guide-for-managing-surveys-interpreting-results-and-influencing-respondents

Jackson, S. L. (2011). *Research methods and statistics: A critical approach* (4th ed.). Cengage Learning.

Kalton, G. (2021). *Introduction to survey sampling* (2nd ed.). Sage Publications Ltd.

Kim, J., & De Dear, R. (2012). Nonlinear relationships between individual IEQ factors and overall workspace satisfaction. *Building and Environment, 49*(1), 33–40. https://doi.org/10.1016/j.buildenv.2011.09.022

Lee, J. W., Kim, D. W., Lee, S. E., & Jeong, J. W. (2020). Development of a building occupant survey system with 3D spatial information. *Sustainability, 12*(23), 9943. http://dx.doi.org/10.3390/su12239943

Li, P., Froese, T. M., & Brager, G. (2018). Post-occupancy evaluation: State-of-the-art analysis and state-of-the-practice review. *Building and Environment, 133*(1), 187–202. http://doi:10.14288/1.0364033

Litwin, M. S. (1995). *How to measure survey reliability and validity*. Sage Publications, Inc.

Mennis, J., Mason, M., & Ambrus, A. (2018). Urban greenspace is associated with reduced psychological stress among adolescents: A geographic ecological momentary assessment (GEMA) analysis of activity space. *Landscape and Urban Planning*, *174*, 1–9. https://doi.org/10.1016/j.landurbplan.2018.02.008

Preiser, W. F. (1995). Post-occupancy evaluation: How to make buildings work better. *Facilities*, *13*(11), 19–28. https://doi.org/10.1108/02632779510097787

Preiser, W. F., Rabinowitz, H. Z., & White, E. T. (1988). *Post-occupancy evaluation* (1st ed.). Routledge.

Punch, K. F. (2003). *Survey research. The basics* (1st ed.). Sage Publications, Ltd.

Wright, K. B. (2005). Researching internet-based populations: Advantages and disadvantages of online survey research, online questionnaire authoring software packages, and web survey services. *Journal of Computer-Mediated Communication*, *10*(3). https://doi.org/10.1111/j.1083-6101.2005.tb00259.x

Xue, P., Mak, C. M., Cheung, H. D., & Chao, J. (2016). Post-occupancy evaluation of sunshades and balconies' effects on luminous comfort through a questionnaire survey. *Building Services Engineering Research and Technology*, *37*(1), 51–65. https://doi.org/10.1177/0143624415596472

Zhang, X., Zhou, S., Kwan, M. P., Su, L., & Lu, J. (2020). Geographic ecological momentary assessment (GEMA) of environmental noise annoyance: The influence of activity context and The daily acoustic environment. *International Journal of Health Geographics*, *19*(1), 50. https://doi.org/10.1186/s12942-020-00246-w

12 Space syntax

Examining human-workplace behavior through isovists and shortest paths

Petros Koutsolampros

Independent Researcher, UK

12.1 Background

Space syntax is a theory and a set of related methods that aim to study the configuration of space and how that configuration links to the behavior of humans inhabiting it. The theoretical base was originally set in a 1976 paper simply titled "Space Syntax" (Hillier et al., 1976) and eventually expanded in The Social Logic of Space by Hillier and Hanson (1984) where a set of methods was also proposed. While these original studies examined mostly small settlements and houses, eventually the domain was expanded to include ways of analyzing spatial structures of different scales such as large urban street networks and buildings of any size. Currently, applications may be found in a variety of fields, from architecture (Peponis et al., 2015) and urban planning (Karimi, 2012) to archaeology (Dawson, 2002), neuroscience (Javadi et al., 2016), and biology (Varoudis et al., 2018). The theory is typically considered applicable to the larger urban scales, primarily due to the substantial number of existing relevant studies; however, there are multiple examples of applications in buildings of different types. These include laboratory buildings (Hillier & Penn, 1991; Serrato & Wineman, 1999; Wineman & Serrato, 1997), factories (Peponis, 1985), hospitals (Pachilova & Sailer, 2020), and schools (Kishimoto & Taguchi, 2014).

At its core, space syntax treats space as a set of parts (rooms, streets) interconnected through permeability or visibility (doors, windows, junctions) forming spatial networks. In large urban forms the result is usually a street network, while in buildings a network of rooms or smaller elements. These networks may then be studied to extract measurements, both about the interconnected parts and the network itself. The measurements are essentially a set of parameters of spatial configuration that both rigorously describe a space (central, isolated, large, elongated) and allow for comparing to human behavior. Statistical analysis of this comparison then allows for understanding how the layout of a space affects that behavior and enables future predictions about new designs.

The first study to consistently apply the methodology for office spaces was by Hillier and Grajewski (1990). This first study directly translated many of the analytical elements of the urban-focused studies that came before, working primarily with lines of visibility and traversal (like streets), rather than areal properties of space such as rooms. Later studies focused on rooms and spaces (Toker & Gray,

2008), or corridors (Kabo et al., 2013), but also brought in direct visibility in the form of properties of space (Appel-Meulenbroek, 2009) or co-visibility between staff members (Beck, 2013). A more expanded theoretical overview of space syntax and its application in workplaces can be found in the study by Sailer and Koutsolampros (2021) as well as in more complete and extensive forms by Sailer (2010) and Koutsolampros (2021).

Applications of the theory typically involve measurements of both a dimension of human behavior (location, activity, perception) and the spatial configuration of a workplace (centrality, openness). The focus on the elements of spatial configuration makes space syntax unique in relation to the larger body of workplace research which revolves more around the dimensions of human behavior. This focus allows for a better understanding of how space affects behavior and how the workplace might be reconfigured to achieve different effects; including optimized space utilization and thus more compact workspaces, and also specific placement of teams to better facilitate collaboration.

12.2   Examining human behavior in relation to spatial layout

The two most common measurements of the human dimension in the space syntax domain have taken the form of either direct observation of the location and activity of staff members, or questionnaires and surveys about members' opinion and perception.

Direct observation specifically refers to the periodic recording of human activity primarily in terms of where it is located, typically captured along with other parameters, such as the kind of activity performed (walking, interacting) or other relevant attributes (staff member team, role). The method was employed by early space syntax workplace studies (Hillier & Grajewski, 1990; Hillier & Penn, 1991) and can be carried out by human observers as described in the Space Syntax Observation Manual (Grajewski, 1992; Vaughan, 2001). This particular form of human activity recording is beneficial to the analysis as it allows for direct spatial comparison against the properties of space the individual was observed in. For example, if a large number of people are found standing in a particular location but not elsewhere, it is possible to statistically examine the size, shape, and centrality of that location to identify whether those spatial parameters relate to the concentration. Section 3.2 discusses the technique applied in a particular project with an example observation shown in Figure 12.4.

The alternative is the more widespread method of asking the staff members to complete a questionnaire, reporting on different parameters depending on the particular study. These parameters can be of a wide variety, from the locations of interactions (Appel-Meulenbroek, 2009; Toker & Gray, 2008) to the less tangible aspects of office culture such as privacy and collaboration (Wineman & Adhya, 2007), productivity (Wineman et al., 2011), and knowledge sharing (Appel-Meulenbroek, 2009). While this method allows for exploring the non-physical aspects of an organization and its community, it makes the comparison to the measurements of spatial configuration, as those are provided by the space syntax methods, harder. In this case a locational parameter (the location of the person responding) must either be provided with the questionnaire (Appel-Meulenbroek, 2009; Sailer et al., 2021) or generated by aggregating to a known spatial parameter, i.e., by grouping a team's responses and matching them to the approximate location of the team within the floor.

These two methods of capturing dimensions of human behavior enabled various comparisons to the properties of workspaces and allowed researchers to identify effects at different scales. Studies examining micro- and meso-scale effects showed how desk positioning affected privacy and openness and thus staff seating preferences (Alavi et al., 2018), how distance between workers influenced interactions and thus knowledge transfer (Allen, 1977), and how the layout affected the office social network (Sailer & McCulloh, 2012). Macro-scale effect studies (involving the complete layout of the building) observed that workspaces influenced serendipitous encounters and in turn innovative outputs such as publications and patents (Wineman et al., 2011). In some cases the outcomes were purely spatial; for example, one study found that the configuration of office space affected the occurrence of movement and interaction, especially when the location of attractors (teapoints, entrance) was taken into account (Koutsolampros, 2021).

12.3   Applying space syntax methods in practice

Space syntax relies on two principal elements for the analysis of buildings, isovists, and shortest paths. An isovist is a representation of the space visible from a particular point, and a shortest path is a route between two points in the building which is the shortest possible either in walking distance or in number of turns. Both elements allow for measuring properties of space, primarily static and positional in the case of isovists, while taking into account the whole layout in the case of paths. The two elements may be used as stand-alone tools, but they also come together under a single method called Visibility Graph Analysis (VGA) where all floor space in the building is divided into square cells (using a lattice grid, typically 50×50 cm) and various metrics are calculated for each of the cells. These metrics include the amount of space visible (i.e., the area of an isovist), the distance of a particular cell in relation to attractors (i.e., the shortest path to the entrance or a teapoint), or the centrality of the cell (i.e., the average number of turns of all paths from that point to all others in the building, usually termed "integration"). Further details about VGA and the various metrics generated can be found in a study by Koutsolampros et al. (2019). All the measurements mentioned above are typically calculated using open-source software such as depthmapX (depthmapX development team, 2019).

Two studies will be used as examples of the above-mentioned elements: (1) an application of single-point isovists representing desk seats aiming to explore perceptions of the workplace environment and (2) a large-scale study of multiple offices employing Visibility Graph Analysis and shortest paths to examine the location of movement and interaction.

The two studies have some base parameters in common. Their datasets both include one or more floor plans describing the buildings examined, obtained from the companies. The floor plans specifically contain elements that may be experienced by occupants (walls, windows, doors, glass, stairs, etc.) and no other construction-related details. The elements contained are made up of simple lines and carry no other properties such as the types of walls, the material of surfaces, or the heights of barriers. In both cases the location of all desks is also known and provided in the floor plans in the form of points.

12.3.1 Open-plan desk positioning and staff perceptions

The first study (Sailer et al., 2021) examined a Tech company in a single office building in London, UK in 2018, and was undertaken by a research team within University College London (UCL) including the author who participated in observation, analysis, and writing. The company was interested in understanding how the density and configuration of desks in their open-plan office affected the perceptions of their staff members. More specifically, the company was trying to understand whether large open plan spaces should be broken up and whether the staff members preferred to have fewer other members around them.

The study employed isovists as a means of examining patterns of co-visibility between staff members and measurements of the immediately visible space from that desk. More specifically, both 360° and 170° isovists were created from each desk (Figures 12.1a and 12.1c, respectively). The software package depthmapX was used for the creation of all isovists through an R package (rdepthmap, Koutsolampros & Kostourou, 2019), leveraging the aforementioned points on each desk in the plan. The isovists allowed for identifying which other desks are potentially visible (a measure termed "degree," Figure 12.1b) and those that are usually visible when someone is sitting at the desk facing forward ("outdegree," Figure 12.1c). Applying the reverse process also made it possible to identify which seats a particular seat is visible from, assuming that everyone is seated. The two isovists further allowed measuring how much space is potentially visible (360°) and usually visible when seated (170°), and consequently to create compound measures such as the density of people visible with regard to the particular isovist's area, and also a ratio of the two areas termed "control."

These spatial parameters were statistically set against a survey distributed to the staff members sitting at the examined desks, to identify how they perceived teamwork, focused work, and productivity in their environment. For example, staff were asked to rate "How much does the workspace at <building> support or inhibit" various elements such as "Accessing coworkers" or "Spontaneous meetings." The response rate was 16%, i.e., 172 answers were collected. The comparison was carried out in R by creating multiple ordinal regression models with the four continuous spatial variables (degree, outdegree, density, control) and seat type acting as independent variables, and each of the survey items acting as the dependent variables. Tenure, industry experience, and role were included as controls.

The models showed that the spatial parameters played a role in the perception of staff members, with the most relevant factors being how many other desks each member permanently had in their line of sight, and how much their back was

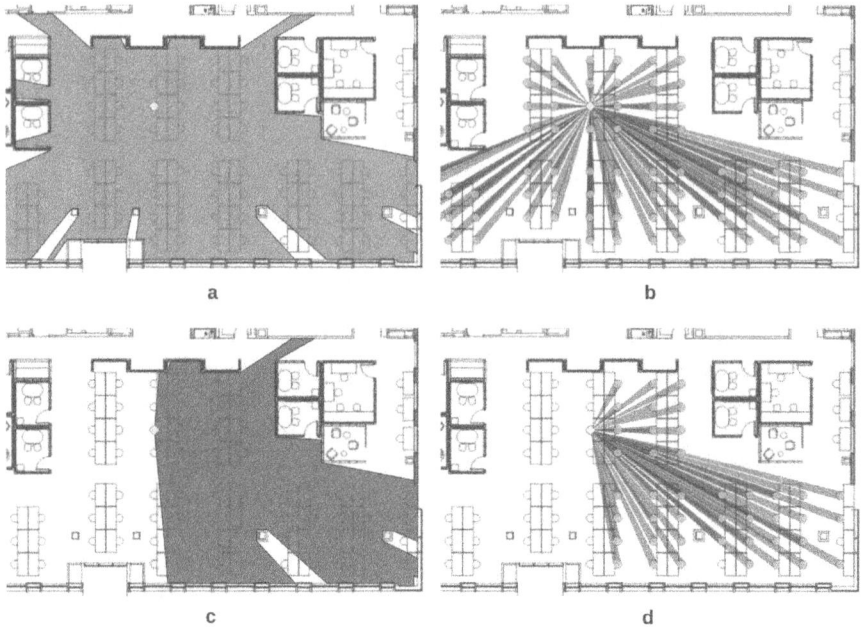

Figure 12.1 Measuring co-visibility and local properties of space from a particular seat
(marked with a yellow rhombus) specifically: (a) 360° isovist from the desk,
(b) all the other seats that might be visible from the particular seat,
(c) 170° isovist taking into account the orientation of the seat, and (d) the seats visible
from the particular seat when the person is seated (Sailer et al., 2021).

protected. More specifically it was found that people with fewer desks in their immediate line of sight when seated (outdegree) rated team identity and cohesion,
concentration, and productive work higher, and that those with a higher ratio of
forward facing area to all their visible space (control) rated sharing information and
team identity and cohesion higher.

   The two findings mentioned were in fact a minority among a set of other hypotheses that could not be statistically confirmed due to low significance of the model
effects. However, the study demonstrated that it is not only possible to measure and
compare spatial parameters provided by space syntax against human perceptions
and opinions, but that more nuance can be added by expanding and clarifying those
spatial parameters. For example, it was shown that it was not only the size of the
visible area that the staff found troublesome but more so the effect of seeing many
others (and thus being exposed to potential distractions).

12.3.2 *Large-scale analysis to identify the locations of movement and interaction*

In a large-scale study for the author's Doctoral Thesis (Koutsolampros, 2021), a
set of 34 companies spread over 60 buildings and 213 floors was examined in
order to understand how the layout affects movement and interaction. The aim of

the research was to develop a complete framework for working with spatial and behavioral data for buildings and more specifically workspaces and to evaluate its effectiveness using the large dataset. In this case the dataset was collected over ten years by Spacelab, a London-based architectural design office, and was provided to the author for analysis. The companies examined were active in various industries, including Retail, Media, and Technology, and of various sizes, from small 30-person workplaces to larger offices with more than 2000 desks.

Prior to any analysis the various floor plans were cleaned up and unified, with additional information added such as the locations of various attractors (toilets, entrances), links between floors and buildings, and demarcation of the various types of spaces such as workspaces, meeting rooms, and circulation.

The study used Visibility Graph Analysis as a method for calculating properties of space, i.e., by dividing the space of each of the buildings into a grid of cells and generating various measurements for each cell as mentioned above. In this case the isovists were used only to calculate local properties of space for each cell, such as the area of the visible space (Isovist Area, Figure 12.2a) and the distance to the closest obstacle (Isovist Minimum Radial, Figure 12.2b). Building-scale properties

Figure 12.2 Properties of space on Visibility Graph Analysis grid: (a) Isovist Area, (b) Isovist Min Radial, (c) Visual Mean Depth, and (d) Metric Mean Depth (Koutsolampros, 2021).

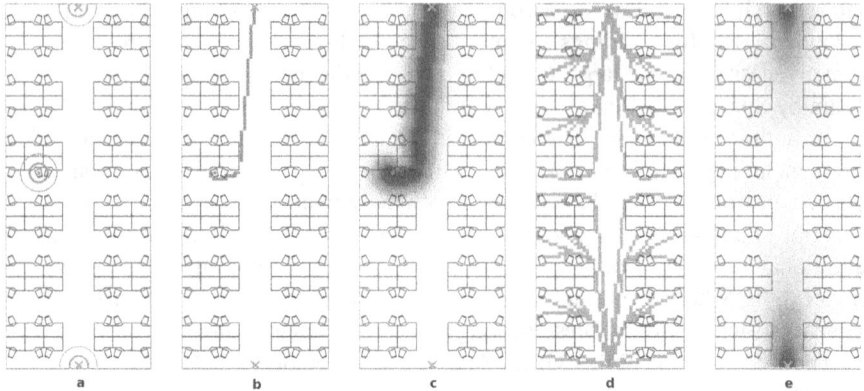

Figure 12.3 Calculation of travel concentration: From every desk choose the closest attractor (a) and create a path to it (b), as well as the path's immediate field (c). Take the same paths from every desk (d) and aggregate them together into a single number for each cell of the grid (e) (Koutsolampros, 2021).

of space were also calculated for each cell, including the average number of turns (Visual Mean Depth, Figure 12.2c) and the average distance (Metric Mean Depth, Figure 12.2d) required to get to the rest of the building.

All the metrics above are known within the space syntax domain, but for this particular study a new metric was developed. Previous research pointed out that local and building-scale properties of space might be insufficient to fully explain human activity, and that the location of attractors should be taken into account. Here this was achieved by taking the shortest paths between each desk and selected attractors (toilets, entrance, teapoints), calculating the immediately accessible area of that path and aggregating on each grid cell to acquire a value of how concentrated the travel is to be expected at that point. The steps of the process are presented in Figures 12.3a–d and the result in Figure 12.3e.

The spatial metrics were compared against hourly observations of human activity which included the location of people in the building, as well as whether they were sitting, standing, walking, or interacting (an example snapshot is shown in Figure 12.4). For the comparison various methods were used, mostly by first aggregating both the spatial and human activity data to areas such as whole floors or rooms and then by carrying out multiple regression analysis. Another method employed was point pattern analysis which allowed for statistical comparisons at the cell level, without aggregation.

This study found strong connections between the examined metrics and human activity (movement and interaction). It was shown that building-scale centrality, taking into account number of turns or walking distance, was an important parameter for movement; however, the expected concentration of travel was the strongest predictor out of the three. Interaction was found to be more complicated, with different metrics affecting different types of interactions, though primarily dominated by the local properties of space.

Sitting
Standing
Walking

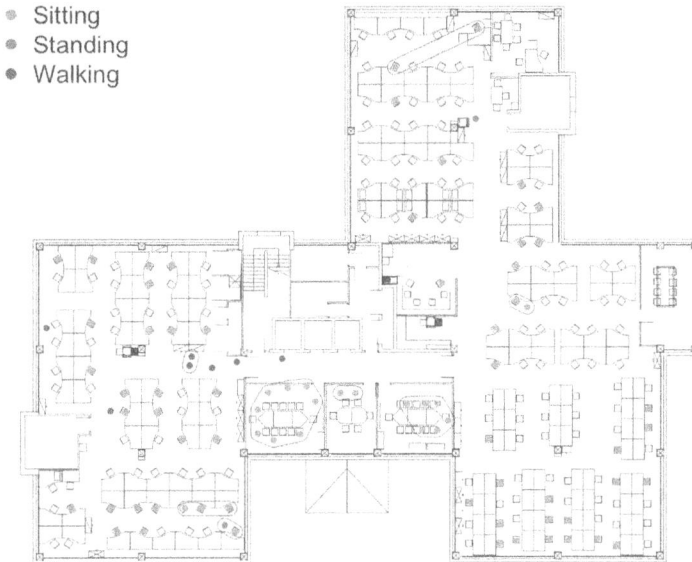

Figure 12.4 Snapshot of an hourly observation of human activity. People sitting as green dots, standing as red, and walking as blue. Encircled dots are people interacting (Koutsolampros, 2021).

In contrast to the earlier example, this study aimed to use existing metrics as much as possible and to explore the relationship between human behavior and spatial configuration more generally. Instead of introducing more nuance, this study employed existing techniques with a large dataset and showed that some of the metrics are already good predictors with a large enough sample.

12.4 Implications

The above-mentioned studies are examples of how to apply the space syntax theory and methods to real-world scenarios as a means of understanding the relationship between workspace and human behavior. Through these examples it is shown that it is possible to link the configuration of space to both tangible (locations of movement and interaction) and intangible (teamwork, focused work perception) human dimensions.

12.4.1 Method relevance to research

The particular methodologies examined are well suited in answering research questions on how space relates to human behavior, primarily because they allow for creating measurable representations of space. Representations such as the isovist help convert visible space to a quantifiable property which may in turn allow for comparing against perceptions of teamwork and productivity, and also opinions about the environment itself such as whether it feels claustrophobic or too open. The isovist also allows for identifying distinct elements that might be visible, such as other seats (as in the example above), greenery, windows, and decoration which eventually allow for examining whether staff members find their environment cluttered or organized, with enough greenery or outside views.

Visibility Graph Analysis and shortest paths, i.e., the representations used in the second example, provide more holistic views of the building and can thus be used to answer questions beyond single locations. In the second example only movement and interaction were examined, but it is also possible to identify the parameters of the layout that most affect occupancy of seats and meeting rooms, as well as utilization of other spaces such as teapoints and informal meeting areas. It is also possible to ask specific questions such as "are central or more visible seats occupied more often than the rest?" or "do more open-plan spaces attract interactions between seated staff members and those who are passing by?"

The two studies also demonstrate both that the existing methodology may provide a strong base for understanding human behavior more generally, and that it can be extended to introduce more nuance if required.

12.4.2 *Method relevance to practice*

The examples above aimed to answer particular research questions, but they were also evaluated using specific case studies. The results are usable as suggestions when designing new spaces; for example, the first study shows that the density of people visible affects focused work and teamwork. Such information might be used by the workplace designer or the company itself to modify a workspace so that staff have the optimum density of people visible that is best for facilitating focused work and teamwork. Using similar frameworks and equivalent or larger samples it would be possible to extract even more specific design strategies such as the optimal locations for various attractors, or how to design a space to maximize teamwork and productivity.

However, perhaps the main potential gain when applying space syntax methodology in practice is for testing and evaluating design decisions. Applying the same methods on the same representations of an existing building and a potential new design will help identify differences in the human dimension. For example, by creating an alternative seating plan for the first example and re-evaluating it, it is possible to project the expected perceptions on teamwork and productivity. In a similar vein it is possible to redesign a building from the second example to increase or decrease instances of movement and interaction so as to achieve quiet or buzzing spaces. In both cases it is then possible to iterate over the design to achieve different outcomes based on the requirements of the organization.

The thesis produced by the second study (Koutsolampros, 2021) briefly demonstrates both applicable potentials. Design suggestions are discussed in the context of a hypothetical design decision, the placement of a staircase (p. 304). Based on the existing results it is theorized, for example, that we can expect movement and movement-related interaction to be high when close to a staircase, but if a new set of stairs is introduced, then those activities can be expected to diminish because some staff members will now be closer to an attractor through a different path. The study also demonstrates how predictions might be made, taking input from the analyzed models both for movement (p. 263) and interaction (p. 273). In this way the models developed might be successful in providing some information about future activity without the need for a direct observation and thus make typical qualitative investigations faster, cheaper, and less disruptive of normal office operations.

12.5 Conclusions

The domain of space syntax contains a rich set of methods and studies for understanding the relationship between spatial configuration and various dimensions of human perception and behavior. It is most suited for cases where the human aspects examined are related to the actual workplace environment, even when those are intangible (teamwork, focused work) as long as they relate to the workspace in some way. Conversely, because space syntax deals with physical space, studies focusing exclusively on intangible aspects of office life have little to gain from its application. Additionally, there appears to be no practical way to apply the theory for companies that work fully remote, and companies that employ hybrid work models are only likely to extract meaningful outcomes for the physical aspects of office life.

In terms of the actual application, the theory may sometimes appear to be complex and inscrutable. The currently available tools are researcher focused and tend to have unintuitive interfaces, thus require spending a lot of time to understand how to use them. There is also a lack of a complete methodological manual, not only owning to the evolving nature of the theory but also because it was typically focused more on the urban scales. Recent studies provide some insights into the complexity, and limited work has been done to make the tools easier; however, larger strides in development can only be expected with wider adoption.

12.6 Further Reading

The reading list provided below focuses on material that takes a general approach in the application of space syntax in workplaces, rather than studies that focus on one particular workplace and technique. The list is provided here with the aim of being a useful entry point for researchers and practitioners.

1 The first suggestion is the study by Hillier and Grajewski (1990) which is the first application of space syntax in office spaces. The study examines a set of buildings using techniques borrowed from the urban side of space syntax and thus acts primarily as a historical reference for understanding the domain in context of workplaces: Hillier and Grajewski (1990). *The application of space syntax to work environments inside buildings: Second phase: Towards a predictive model* (Unit for Architectural Studies). The Bartlett School of Architecture and Planning, University College London.

2   Following is the doctoral thesis by Sailer (2010) which takes a holistic approach to the examination of spatial configuration and organisational behaviors, bringing together various methods and techniques that were developed within the domain until that point. The thesis not only contains an extensive historical overview of both organisational behaviors (as they relate to space) and the various elements of space syntax but also demonstrates how to apply the methods on a set of organisations: Sailer (2010). *The space-organisation relationship. On the shape of the relationship between spatial configuration and collective organisational behaviours* [Doctoral dissertation]. TU Dresden.

3   The author's own doctoral thesis (briefly examined as the second example in Section 3.2) is also referenced here, as a continuation of Sailer's work, but focused on fewer methods and an expanded sample of workplaces. The thesis only uses (and expands on) Visibility Graph Analysis for measurements of spatial configuration compared against direct observations of human behavior but applies the methodology across a sample of 34 companies. The thesis also contains a review concentrating specifically on the models and datasets used through the years: Koutsolampros (2021). *Human behaviour in office environments— Finding patterns of activity and spatial configuration in large workplace datasets.* [Doctoral dissertation]. University College London.

4   The only single-workplace study suggested is the one examined as the first example in Section 3.1. It is referenced here to show that it is possible to expanding the techniques found within the domain of space syntax to address specific questions. The use of isovists allowed the authors to rigorously measure aspects of perception (amount of visible space, number of others visible to) and then compare them to related opinions (on team identity and cohesion and number of distractions): Sailer and Koutsolampros (2021). Space Syntax: Understanding human movement, co-presence and encounters in relation to the spatial structure of workplaces. In *Theories for Achieving Fit between People and the Office Environment: A Transdisciplinary Handbook.* Routledge.

5   The final item referenced is the only currently available manual describing the various techniques and methods used in the space syntax domain in a succinct manner and not related to a particular study. The manual contains both urban and building-scale techniques without focusing on particular cities or building types: Al-Sayed et al. (2014). *Space syntax methodology.* Bartlett School of Architecture, UCL. http://discovery.ucl.ac.uk/1415080/

References

Alavi, H. S., Verma, H., Mlynar, J., & Lalanne, D. (2018, April). *The Hide and Seek of Workspace: Towards Human-Centric Sustainable Architecture.* Proceedings of the 2018 CHI Conference on Human Factors in Computing Systems, Montreal, Quebec, Canada. https://doi.org/10.1145/3173574.3173649

Allen, T. J. (1977). *Managing the flow of technology: Technology transfer and the dissemination of technological information within the r & d organization.* MIT Press.

Al-Sayed, K., Turner, A., Hillier, B., Iida, S., & Penn, A. (2014). *Space syntax methodology.* Bartlett School of Architecture, UCL. http://discovery.ucl.ac.uk/1415080/

Appel-Meulenbroek, R. (2009). Knowledge sharing in research buildings and about their design. In D. Koch, L. Marcus & J. Steen (Eds.). *Proceedings of the 7th international space syntax symposium* (pp. 004:1–12). Royal Institute of Technology (KTH): Stockholm, Sweden. http://www.sss7.org/Proceedings/04%20Building%20Morphology%20 and%20Emergent%20Performativity/004_AppelMeulenbroek.pdf

Beck, M. P. (2013). Visibility and exposure in workspaces. In O. Kim, H. T. Park & K. W. Seo (Eds.), *Proceedings of the 9th international space syntax symposium. 9th international space syntax symposium* (pp. 017:1–10). Sejong University, Seoul, Korea http:// sss9sejong.or.kr/paperpdf/bmp/SSS9_2013_REF017_P.pdf

Dawson, P. C. (2002). Space syntax analysis of Central inuit snow houses. *Journal of Anthropological Archaeology*, *21*(4), 464–480. https://doi.org/10.1016/s0278-4165(02)00009-0

depthmapX development team. (2019). *DepthmapX* (0.7.0). https://github.com/SpaceGroup UCL/depthmapX/

Grajewski, T. R. (1992). *Space Syntax Observation Manual*.

Hillier, B., & Grajewski, T. R. (1990). *The application of space syntax to work environments inside buildings: Second phase: Towards a predictive model* (Unit for Architectural Studies). The Bartlett School of Architecture and Planning, University College London.

Hillier, B., & Hanson, J. (1984). *The social logic of space*. Cambridge University Press (CUP). http://dx.doi.org/10.1017/cbo9780511597237

Hillier, B., Leaman, A., Stansall, P., & Bedford, M. (1976). Space syntax. *Environment and Planning B: Planning and Design*, *3*(2), 147–185. https://doi.org/10.1068/b030147

Hillier, B., & Penn, A. (1991). Visible colleges: Structure and randomness in the place of discovery. *Science in Context*, *4*(1), 23–50. https://doi.org/10.1017/S0269889700000144

Javadi, A.-H., Emo, B., Howard, L., Zisch, F., Yu, Y., Knight, R., Pinelo Silva, J., & Spiers, H. J. (2016). Hippocampal and prefrontal processing of network topology to simulate the future. *Nature Communications*, *8*, 14652. https://doi.org/10.1038/ncomms14652

Kabo, F., Hwang, Y., Levenstein, M. C., & Owen-Smith, J. (2013). Zone Overlap and Collaboration in Academic Biomedicine: A Functional Proximity Approach to Socio-Spatial Network Analysis. *Ross School of Business Paper, 1184*.

Karimi, K. (2012). A configurational approach to analytical urban design: 'Space syntax' methodology. *Urban Design International*, *17*(4), 297–318. https://doi.org/10.1057/udi. 2012.19

Kishimoto, T., & Taguchi, M. (2014). Spatial configuration of Japanese elementary schools: Analyses by the space syntax and evaluation by school teachers. *Journal of Asian Architecture and Building Engineering*, *13*(2), 373–380.

Koutsolampros, P., & Kostourou, F. (2019). *rdepthmap: R and depthmapX CLI interface* (0.1.0). https://github.com/pklampros/rdepthmap

Koutsolampros, P., Sailer, K., Varoudis, T., & Haslem, R. (2019). Dissecting visibility graph analysis: The metrics and their role in understanding workplace human behaviour. *Proceedings of the 12th International Space Syntax Symposium*, 24. https://discovery.ucl. ac.uk/id/eprint/10073528/

Koutsolampros, P. (2021). *Human behaviour in office environments—Finding patterns of activity and spatial configuration in large workplace datasets* [Doctoral dissertation]. University College London.

Pachilova, R., & Sailer, K. (2020). Providing care quality by design: A new measure to assess hospital ward layouts. *The Journal of Architecture*, *25*(2), 186–202. https://doi.org/ 10.1080/13602365.2020.1733802

Peponis, J. (1985). The spatial culture of factories. *Human Relations*, *38*(4), 357–390. https://doi.org/10.1177/001872678503800405

Peponis, J., Bafna, S., Dahabreh, S. M., & Dogan, F. (2015). Configurational meaning and conceptual shifts in design. *The Journal of Architecture*, *20*(2), 215–243.

Sailer, K. (2010). *The space-organisation relationship. On the shape of the relationship between spatial configuration and collective organisational behaviours* [Doctoral dissertation]. TU Dresden.

Sailer, K., & Koutsolampros, P. (2021). Space syntax: Understanding human movement, co-presence and encounters in relation to the spatial structure of workplaces. In. In R. Appel-Meulenbroek, & V. Danivska (Eds.), *A handbook of theories on designing alignment between people and the office environment* (pp. 248–260). Routledge.

Sailer, K., Koutsolampros, P., & Pachilova, R. (2021). Differential perceptions of teamwork, focused work and perceived productivity as an effect of desk characteristics within a workplace layout. *PLOS ONE*, *16*(4), e0250058. https://doi.org/10.1371/journal.pone.0250058

Sailer, K., & McCulloh, I. (2012). Social networks and spatial configuration—How office layouts drive social interaction. *Social Networks*, *34*(1), 47–58. https://doi.org/10.1016/j.socnet.2011.05.005

Serrato, M., & Wineman, J. D. (1999). Spatial and Communication Patterns in Research & Development Facilities. *Proceedings 2nd International Space Syntax Symposium*. 2nd International Space Syntax Symposium, Brasília, Brazil. http://www.umich.edu/~igri/publications/SpaceSyntax.pdf

Toker, U., & Gray, D. O. (2008). Innovation spaces: Workspace planning and innovation in u.S. University research centers. *Research Policy*, *37*(2), 309–329. https://doi.org/10.1016/j.respol.2007.09.006

Varoudis, T., Swenson, A. G., Kirkton, S. D., & Waters, J. S. (2018). Exploring nest structures of acorn dwelling ants with x-ray microtomography and surface-based three-dimensional visibility graph analysis. *Philosophical Transactions of the Royal Society B: Biological Sciences*, *373*(1753). https://doi.org/10.1098/rstb.2017.0237

Vaughan, L. (2001). *Space syntax observation manual (2001 unpublished revised edition)*. Space Syntax Ltd.

Wineman, J. D., & Adhya, A. (2007). Enhancing workspace performance. *Proceedings of the 6th International Space Syntax Symposium*, *66*, 1–16. http://www.spacesyntaxistanbul.itu.edu.tr/papers/longpapers/066%20-%20Wineman%20Adhya.pdf

Wineman, J. D., Kabo, F., Owen-Smith, J., & Davis, G. (2011). Spatial layout and the promotion of innovation in organizations. *Considering Research: Reflecting Upon Current Themes in Architectural Research*, *327*.

Wineman, J. D., & Serrato, M. (1997). *Enhancing Communication in Lab-based Organisations*. Proceedings of the First Space Syntax International Symposium, Seoul, South Korea.

13 Journey mapping

Describing the spatial experience of workplace users

Antonio Iadarola

Studio Wé, USA

13.1 Background

With the increased flexibility in work arrangements, employees are exposed to a complex set of interactions in different personal spaces and work spaces and at different times in their daily routines. Designing workplaces for this higher degree of flexibility deserves more attention as it can challenge the quality of workers' satisfaction, productivity and well-being. This chapter introduces the application of experience journey mapping (JM) in the context of workplace design and management to analyse and design workplace experiences.

To approach this method we define an **experience** as "the way that something happens and how it makes you feel" (OED, 2013).

In general, JM visualises all the phases and interactions of a user experience with a product, service or environment. JM is an established method in human-centric design disciplines like User Experience (UX) and Service Design to design multi-channel product-services. In User Experience Design the term channel indicates the medium of interactions between a customer and an organisation, product or service (e.g. person-to-person, printed, spatial, online, mobile, etc.) The output of a JM activity is a journey map: a schematic, typically matrix-like visual document, used as a proxy to represent one or more user journeys of interacting with a product, service or environment. Multiple authors speak about designing the "employee experience" (Bridger & Gannaway, 2021; Whitter & Bersin, 2019; Wride et al., 2017), namely: looking at how employees access internal services and organisational resources to support their development and work functions. Consultancy companies as well often talk about employee or workplace experience (Baird et al., 2022; Carrión, 2021; Deloitte, 2021; Gallup, 2022).

Workplace experience is the result of the use of the work environment over time (Leesman, 2018). However, the use of JM to manage such experiences and add a temporal understanding of worker interactions, is rarely mentioned in workplace research (Fassi, 2018a,b), and it has no clear definition in this context. The idea of visualising an experience across multiple touchpoints – artefacts that allow interaction between the user and a service (e.g. an interface, a form, a space, a person offering support) – originated from Jan Carlzon's concept of "moments of truth":

DOI: 10.1201/9781003289845-13

Table 13.1 Types of journey maps, adapted from chapter "Mapping experiences," Kalbach (2016, p. 11)

Mental model diagram	*Experience map*	*Customer journey map*	*Service blueprint*	*Spatial map*
Broad exploration of human behaviours, feelings, and motivation	Illustrates experiences of people within a given domain (e.g. travelling, shopping, banking, etc.)	Illustrates the experience of an individual as a customer of an organisation	Diagram of a service offer, visualising in parallel customer actions and business processes	Diagrams mapping out an experience spatially

moments when users form an impression of a product, brand, or service that influences their behaviours and decision-making patterns (Carlzon, 1987).

The method, called Customer Journey was popularised in business management in the late 1980s and was immediately adopted in software development and marketing in the 90s as "a pictorial representation of the experience clues to be engineered" (Carbone & Haeckel, 1994). Since then, JM has become widely used by designers and managers for product-service development and strategy in any industry. In the last two decades, many typologies of experience maps were developed, mostly through practice-based iterations in service organisations and design agencies. Jim Kalbach summarises them in Mapping Experiences (Kalbach, 2016), the most comprehensive volume on the topic (Table 13.1).

Kalbach emphasises that all these experience visualisations are important alignment tools between providers' intentions and user needs. Achieving that alignment ensures value creation for both parties (Kalbach, 2016). In the workplace design context, these tools effectively align organisations' strategic goals with workers' needs. In the context of Workplace Design (WD), JM can support the advancement of workplace research and practice by introducing a qualitative approach to understand subjective workplace experiences grouped by user types. Integrated with quantitative data, JM enriches workplace analysis and evidence-based decision-making. Design practitioners and managers can use JM to look at the workplace from the privileged perspective of employees and customers that can directly experience physical spaces, digital tools, as well as processes, policies, and rituals within organisations.

13.2   Argument

Experiential environments that provide a mix of sensorial stimuli and possibilities to modify and adapt the space based on individual and team preferences, increase workers creativity and capacity to innovate (Sanders & Stappers, 2012). Better communication, facilitated collaboration and engagement with spatial services (e.g., coworking or corporate facilities) are also results of an intentionally designed experience (Chafi & Cobaleda Cordero Cobelada, 2021; Cobaleda-Cordero & Babapour, 2022). They are factors contributing to organisation's success (Manca et al., 2018). JM offers an effective tool and process to orchestrate the diversity of these elements. However, its application for spatial-based, particularly workplace experiences, is still sparse compared to UX and service design, where JM is an essential method. This chapter will present examples of its favourable adoption by comparing and contrasting the method in service-based and spatial experiences (Groves & Marlow, 2016; Marlow & Groves, 2013). Usually spatial experiences and service-based experiences are documented respectively by means of spatial and non-spatial maps.

Spatial maps visually indicate the positions and proximity of the physical elements within the space represented. For example, the spatial Data Mapping created for Thomas Jefferson University Hospital by architect KieranTimberlake (see Figure 13.1) explores how space is used in the emergency department and shows occupancy data of different users in the hospital. In spatial maps, relations between the built environment and the emotional state and cognitive conditions of the users are hard to represent visually, as well as the relations between the space and its social ecosystem, making it harder to locate information that generates insights on the user experience.

On the other hand, **non-spatial maps** (see Figure 13.2) present information by describing chronological phases of how users experience events. In these maps, user research findings are typically organised schematically instead of represented on the space floorplan. Namely, they allow to represent and analyse, in a simplified manner, complex sequences of interactions over an extended period of time and their users' "pain points," meaning any step or situation during the experience

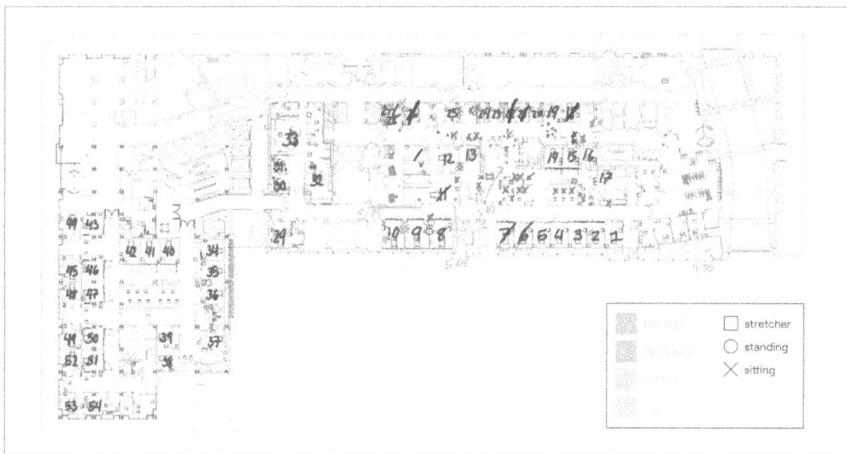

Figure 13.1 Adapted by the author from the book *Health Design Thinking* (Ku & Lupton, 2022). Spatial Data Mapping created for Thomas Jefferson University Hospital by architects KieranTimberlake.

Discover community   Overwhelmed with information   Can't find provider   Wrong Pr... Referr... Somewhe...

EMOTIONAL LEGEND

- High Energy Negative — Anger, overwhelmed, frustrated, anxious, afraid
- Low Energy Negative — Sad, depressed, apathetic, exhausted
- Low Energy Positive — Content, calm, accepting, relaxed
- High Energy Positive — Happy, confident, empowered, surprised, reassured

Recognize that there is an issue	Seek out information on the issue	Learn more about issue	Evaluate options based on what to do	Make a decision on health providers	Book & Travel to health care Service(s)	Access care ser...

Q AWARENESS AND LEARNING	→ MAKING THE DECISION	+ ACC...

Jacqueline, 61 "Fight for your Right"	She searches his symptoms on Google; initial symptom searches point to diabetes	She conducts further research on WebMD, Mayo Clinic, and health science websites	Her husband has an episode where he almost passes out; she calls Telehealth who advises her that he should drink water and rest.	In hindsight she should have taken him to the emergency; Telehealth's advice was dangerous. She decides to contact the family doctor.	She books an appointment with the family physician in 2 weeks	Visiting the ... doctor, her f... expresses h... feeling 'com... comme ça', ... cannot find ... precise wor... English
Jacqueline's husband is not feeling well, and she noticed his demeanor has changed.						

Chris, 23 "Advocate by Day"	Calls his paramedic friends and sends a picture of the bump - they determine it is a wort.	Goes online to learn more about treatment.	Joseph does not have a family physician, so Chris advises he'll drive him to the clinic next day when they are open.	Joseph obsesses over the wart and cannot wait for the clinic - he tries to remove it in his room with a knife, and calls Chris with an open wound.	Chris wraps him up and brings Joseph to the emergency	The emerge... entrance ha... oriented so ... while to find
Chris is notified that Joseph, a folk in the group home at which he works, has a bump on his knee. Joseph is convinced it is something terrible.						

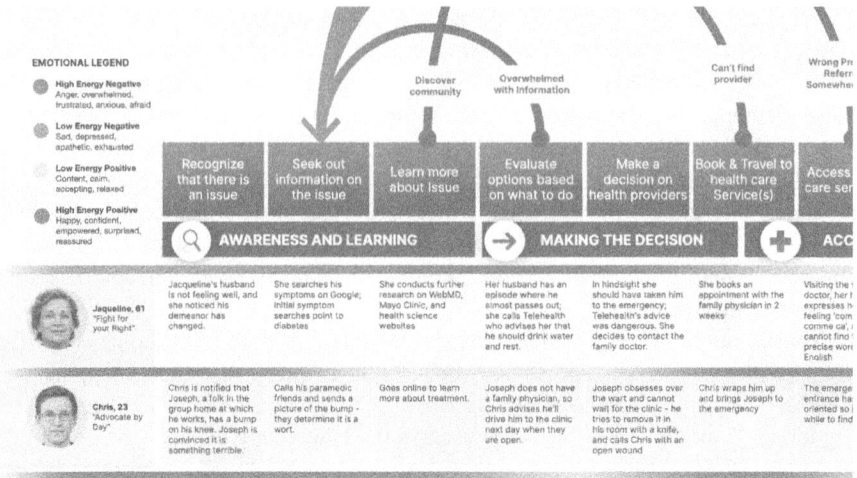

Figure 13.2 Detail of a journey map created by Studio Wé for a healthcare experience project. The top bar and arrows explain the phases of the experience and the horizontal red orange and yellow bars show the level of satisfaction of different user groups.

where the user faces a problem, a frustration or a blocker to progress in achieving their goals. JM may be usefully represented with a matrix where the top horizontal row describes a chronology of phases and sub-phases of the experience and the rows below (also called *swimlanes*) contain insights and information on the user interactions, emotion and thinking processes at each stage of the experience, divided by channel. The hierarchy of the rows is customisable to contain the most suitable information and any analysis criteria for the purpose of the project, as well as a look and feel conducive to understanding the key information. Any pain point is pinpointed in the map. Nevertheless, non-spatial maps usually remain at a high level of abstraction and lack representation of the physical space. This prevents from: (1) easily annotate what characteristics of the space affect the user experience; (2) map different elements and categories of information at different zoom levels directly on a spatial representation.

Therefore, the key difference between the two representations is that spatial maps naturally allow detailed representation of the built environment while lacking room to include the temporal representation of the user experience across the physical space. Vice versa, non-spatial maps appropriately represent sequence of multi-channel experiences with details on each interaction, while lacking visual representations of spatial characteristics affecting the user experience. A comparison of the two approaches is provided in Table 13.2.

From this simple observation is intuitive that the best way of JM for the workplace is to combine the spatial and experiential elements of the two representations, to fully understand the interplay between the actual physical space and its multi-channel user experience. In this chapter, I will show how it is possible to combine

Table 13.2 This table shows the discrepancies in focus and visualisation found between non-spatial and spatial journey maps

Non-spatial journey mapping	Spatial journey mapping
Chronological	Based on areas and layers
Hierarchical	Data distributed onto space visualisation
View of experience of one user persona	Can include experiences of multiple actors in multiple areas
Sequential and goal-oriented	Activity type-oriented
Describe specific outcomes for each experience phase	Describe flows through the space and experience of activity areas
Aims at task effectiveness	Aims at a positive sensorial and cognitive experience of the space
Synthesise experience as a step-by-step process	Synthesise experience as clustering of activities in space
Visualised in matrix-like format	Visualised either on a floorplan, a draft or a conceptual drawing of a space
Matrix format lends itself to be co-created through a workshop session	Co-creation requires more preparation of visual and 3D elements
Focus on the user's thinking processes and feelings	Focus on actors' movement and collaborative use of the space

the two mapping approaches with a case study to describe key phases of the method application. In the following sections I will often adopt the broader term of "actor" to include all the participants "in an action process" (Oxford Dictionary, 2013). Actors hold influence over what they are interacting with. The term actor can also allow to extend past the limits of human-centred design and include objects and tools as having agency in a workspace process, as defined by Bruno Latour's (2005) actor-network theory (Law & Singleton, 2013). It is important to note that "the map is not the territory" (Korzybski, 1933): JM helps to organise categories of information to inform actions and initiatives of experience and service improvement. But, as obvious at is sounds, the completion of the map is not enough to automatically provide implementation of solutions in the complex context of an organisation.

The intrinsic value of JM lies in its process that enables various actors to activate better experiences through different degrees of collaborative activities.

13.3   Example of application/use

JM synthesises in visual outcomes insights from qualitative research techniques like semi-structured user interviews, user testing, facilitation of co-creation workshops, and secondary desk research. This necessary expertise is provided by multi-disciplinary professionals from UX & service design, ethnography, organisational design, social sciences, and business strategy amongst others.

Intuitively, a journey map can be created with very traditional supports like paper and sticky notes on the wall (ideal for in-person workshops) or with digital tools (e.g. a spreadsheet, digital whiteboards like Mural) and dedicated software (e.g. Smaply, or graphic design software).

Whatever the supporting tool, it usually develops into consequential steps/activities that can be summarised as follows based on well-known industry best practices and backed by abundant literature in the field of experience and service design (e.g. Carter, 2022; Downe, 2020; Lichaw, 2016; Stickdorn et al., 2018 amongst others).

Here I will provide a theoretical overview of the recommended phases for successful experience mapping and design before looking at an example focusing specifically on the application of JM in a workplace transformation project.

13.3.1 Steps to create a journey map

1 **Scope the mapping effort:** define what experience you are mapping, from the point of view of which user type, and whether it is more relevant for the project to map the current state (analysis of how the experience currently works) or future state (how would the ideal experience be in the future). Ideally these decisions are taken collaboratively with all actors involved, through co-creation activities like work sessions with key project decision-makers and workshop with the client team following the project to specify primary and secondary goals of the project and share existing information and documentation functional on the organisation's internal processes.

2 **Conduct user research:** user research aims at collecting qualitative and quantitative data on user behaviours and needs, as a base of information to synthesise in the journey maps. User research most common technique is semi-structured users interviews, in this format, of interview designers set important topics to discuss and engage in open conversations with users on the experience to be mapped. In this way, users can share their personal perspective on a certain experience, product or service. The semi-structured nature of the interview script is ideal for JM as it allows users to contribute information on new aspects of their experience yet to be known by the interviewer. Other techniques that usually are part of user research are contextual observation, in which researchers spend time in the space with users and providers to understand their tasks, interactions and processes, eventually asking questions on what they observed. Finally, discovery workshops with internal stakeholders are also a common technique to analyse operational processes and the provider perspective of interactions that users describe in the interviews. To move from the complexity of all the information collected to usable pieces of information directly functional to JM, one common result of user research is the creation of "personas." "Personas" are fictional characters informed by user research, representing different needs, behaviours and goals towards a product, service or environment. A journey map represents an experience from the point of view of a specific persona, as different personas (representing different user behavioural patterns) might experience and use the same service or product in different ways. Therefore user research and the consideration of all results from the techniques applied are indispensable to ground the map into reality and make it a guide for change later in the design stage;

3 **Co-create the first version of the map:** organise user research findings in a visual structure that fits the focus of the mapping effort and highlight pain points. Examples of prompt questions to inform the structure of the map are in line with the scoping phase: What phases of the experience you are analysing? What channels are in scope? Are there other criteria to be mapped? What timeframe of the experience are you mapping? Ideally, the map format should also be co-created with all actors so that they can contribute insights and information while creating the map in facilitated sessions.
4 **Extract and test design solutions:** collaboratively with selected groups of stakeholders look at the map to summarise visible pain points and define opportunities for change and redesign. This helps groups to reach common understanding of complex multi-factor challenges and where to focus resources.
5 **Storytelling and activation:** finally, share the results with actors that were not involved in the map creation and ask them for feedback on the pain points and opportunities identified. Determine key collaborators in your project and invite them to update the map with new research findings over time.

13.3.2 Project for the new branches of a commercial bank by the design consultancy frog

13.3.2.1 Goals

The goal of this project was to change the way banking services were offered in the branches of a commercial bank. The project was led by the global creative consultancy frog (www.frog.co). The bank wanted to reorganise the employees' way of working and renew their customer experience to innovate banking services. The new branch model envisioned included more possibility for self-service transaction, emphasis on digital banking and reduction of fixed seats and the introduction of standing "head-on" work stations, multi-use standing desks where advisors could face entering customers to welcome and support them, as opposite or private enclosed work stations.

We will show how the five JM phases described above were followed and use the case study to clarify additional practical directions for the application of JM specifically in workplace projects.

13.3.2.2 Process

1 **Scope the mapping effort:** In this phase frog scoped the strategic objectives through work sessions and client workshops that crucially identified the objectives for the new user experience of the branches and the elements to be mapped and redesigned:

  1 As a shift from a traditional in-branch experience, where clients and cashier are separated through a counter, the client wanted to provide a more personalised experience where employees have space and time to listen to clients without barriers to propose commercial baking solutions;

2 To allow this more personalised experience, customers in need of cashier operations (e.g. withdraw or deposit money, checking account status, etc.) had to be redirected to self-service in-branch systems and online banking so that the in-branch advisors could focus on consulting customers;

3 The scoping phase also allowed to consider the real estate point of view, in which the reorganisation had create space for new services, like real estate agents and locker, and, above all, inform a flexible, scalable, and adaptable format for all branches, with and unify and optimise them and their cost of implementation.

In workplace projects, at this stage, the scope should be very open to gather any type of organisational needs. However, there are specific channels of the user and employee experience that are constant in physical workspaces and were observed closely in this case study too: (1) Interactions within the physical space; (2) Interactions with digital tools; (3) Human-interactions: collaborative work processes and synchronous-asynchronous communication; (4) Interaction with the workplace organisation: employee recruitment and training, continuous learning, and HR functions. Each of these channels directly informs the design of user-facing and management tools including: spatial configurations and settings; digital tools for communication and collaboration; Employees' playbooks and work methodologies, and more.

2 **Conduct user research:** frog conducted 60+ hours of interviews with clients and employees to analyse user needs. As described in the recommended steps for JM, in this phase the frog team engaged in contextual user interviews, which in this case was conducted by interviewing users directly when visiting the branch to ask question about their experience, understanding of the service offered and online/in-person banking habits. By triangulating user responses around the above-mentioned topics with the collection of direct observation of the user visits in the branch, and the background information developed in the scoping phase, was possible to differentiate user personas and allowed to categorise different "use cases" depending on tasks to accomplish, personal preferences and the different purposes for customers to come to visit the branch in person. Some of the use cases identified where: "Client with appointment," that use the new ATM interface machine and interface for the first time, and goes in-branch to speak with a wealth management advisor; "Client without appointment," that enters the branch to open a new account; "Client for cashier operation," that goes to the branch to complete a transaction with the cashier.

3 **Co-create the first version of the map:** the research findings discovered in the previous phase through user research provided the necessary content to start describing the current experience in a journey map template, and then directly map it onto the floorplan of the branch selected for the pilot, to inform ideation of the new version of the branch experience.

In this work in progress version, the map was presented to users and client to allow co-creation of the future state of the experience. In particular we can see how, starting from a simple sketch of the space provided to the participants in a workshop, the user flows and personas were collaboratively represented onto the floorplan map. This technique, commonly called "desktop walkthrough" lend it-self perfectly to describe spatial-based experiences to the stakeholders involved in the process, and is a very generative starting point for more refined represen-tation of both the space and other tangible or digital touchpoints to be designed.

The new customer experiences co-designed in this phase described the new branch revolving around a living area, visible from the branch's window and the ATM area always located at the entrance. The client comfortably flows through the living area while engaging with the welcoming representatives. They iden-tify their reason to visit the branch and orient them to either the self-service/ATM area or the appropriate advisors. The advisor meetings happen in open or closed nooks located around the main space, depending on the degree of privacy required by the specific customer need. Non-visible from the living area there are staff-only working desks for back office and administrative tasks.

Directly matching the visualisation of the physical space with the description of the current and intended future customer experience, immediately helped un-derstanding spatial constraints and opportunities for change.

4 **Extract and test design solutions:** this phase included 70 hours of collaborative workshops to map the services and describe types of interactions per area. Using the map containing information about user insights and flows not only helped stakeholders fully understand their customer needs, but also provided a clear visu-alisation to move from the current state of the experience to possible future solu-tions. At this point of the project the journey maps completed so far were used to consider new concrete solutions on how to reorganise the space and the staff positions for the new service offer. This exercise resulted in a number of opportu-nities. The map helped prioritising what to prototype, by comparing the solution to the user needs and organisational goals identified through user research. Con-sequently, it made moving fast to "prototyping" (Phases 3 and 4) easier and ef-fectively communicates the new design to stakeholders and the architecture team.

Interestingly there is an evolution of visualisation techniques compared to matrix-like conventional maps and the sketch version with user flows. At this point, the spatial journey map was iterated in a more schematic format to indi-cate which element of the experience to test in each area, and who is in charge of managing it in the client organisation (Figure 13.3).

At this stage of the project we can see how JM evolved and became a blue-print for testing new interactions in the physical space with different type of users. A service blueprint is a diagram of a service offer, visualising in parallel customer actions and provider's front-end and back-end processes. For example in Figure 13.4 we see how the experience of "client with appointment" was mapped describing the sequence of their actions in different dedicated zones of the floorplan.

The spatial-based journey map of the user testing was paired with a more tra-ditional format (Figure 13.5) that is functional to describe the staff interactions in each phase of the user's branch visit.

5 **Storytelling and activation:** the collaborative workshops to map the services interactions per area of the previous phase, led to a 1:1 prototype of the new

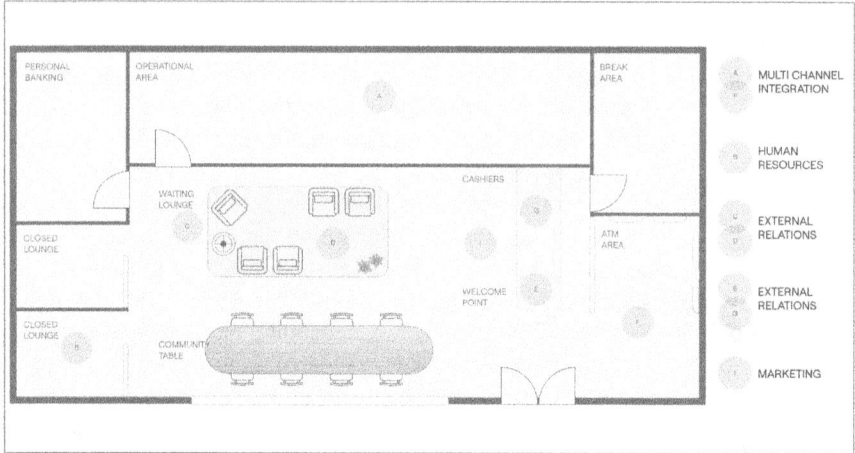

Figure 13.3 Adapted by the author from frog's project documentation. Example of layout of the new pilot space with description of area functions and tags for different teams involved in service implementation.

branch. Over 600 clients and 300 employees engaged in testing the full-scale space prototype, as the design was refined in real-time to better serve their needs.

In this phase, constantly updating versions of JM formats, shown in Figures 13.3, 13.4, and 13.5, became a system to intake and collect insights and data from user testing. This living documentation became crucial to carry all the learnings from the pilot to the iteration of the service model in other branches. Visual communication design to support in-branch interactions and wayfinding to indicate new services were added to the prototypes.

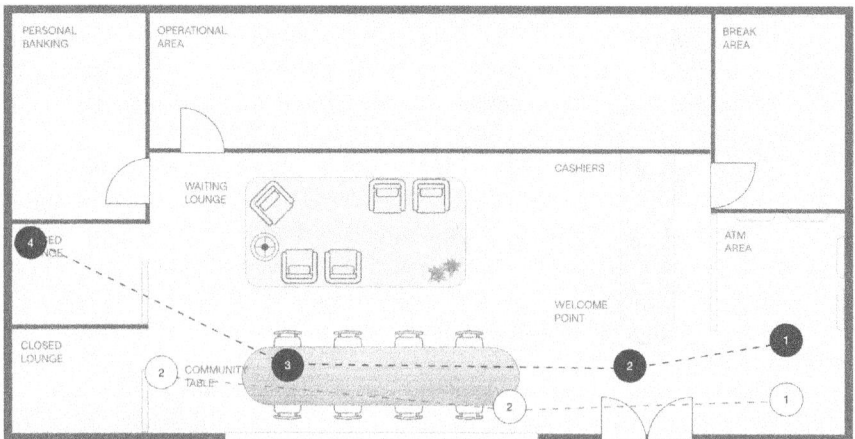

Figure 13.4 Adapted by the author from frog's project documentation. Example of journey map for user persona "client with appointment."

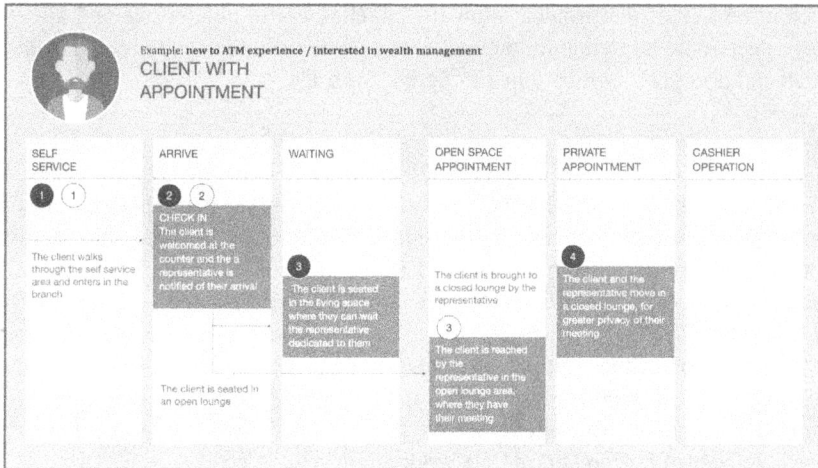

Example: new to ATM experience / interested in wealth management
CLIENT WITH
APPOINTMENT

| SELF SERVICE | ARRIVE | WAITING | OPEN SPACE APPOINTMENT | PRIVATE APPOINTMENT | CASHIER OPERATION |

Figure 13.5 Adapted by the author from frog's project documentation. Example of journey map for user persona "client with appointment" that describes the chronological experience in tandem with the interactions and path of user in space.

In this phase, layering details in a number of new versions of the journey map became a way to have a central source of specifications of the multi-channel experience, which typically includes a diverse set of touchpoints. Pilots were launched in three locations, located in three different cities, to observe the new design in action and refine operations in real scenarios. The company had an in-house architectural team that modelled the branch from an architectural point of view and worked alongside the team of service designers to redesign and map the experience on the space.

Eventually, in the words of the Head of Design of the client, *"The result is a new branch design that fundamentally reframes customer interactions with the bank."* The new design was later extended to all the bank branches around the client's country.

In this example, the spatial journey maps directly informed the five weeks of testing and rapid iterations in a full-scale branch prototype. After less than one year from the project started the bank was able to launch 50 branches based on the service model represented in refined versions of the journey map. Based on this case study we see how spatial-based JM can function as a guideline for a radical workplace redesign and its adaptation in multiple locations.

13.4 Implications

In workplace design and management, JM acts as a translator between business decision-makers and workplace managers and designers by enabling evidence-based workplace decisions. The different versions and visualisations of JM that are shown above informed strategic decisions, and facilitated connection between different phases of the project, allowing stakeholders to collaborate and align on design decisions, implementation, and care of the workplace experience. Both research and practical implications can derive from JM.

13.4.1 Method relevance to research

In academic research, JM can be applied to gain a multi-actor perspective of a workplace ecosystem and understand the human-centric view of the organisational context. Because the insights contained in a JM come from user research on a specific context (e.g., the user experience of a specific physical environment), they might not be suitable to generalise research conclusions valid for other contexts, yet the collection of these subjective views, in parallel with hard data coming from real estate analysis, results in a complex and holistic perspective that is a privileged standpoint to understand contemporary socio-technical phenomena like, for example, the impact of hybrid and remote work.

13.4.2 Method relevance to practice

The case study contributors articulated two types of impact that JM had on their workplace design and management.

First, JM allowed frog's design team to shape new spatial functions and way of working around revenue-generating activities of the organisation. The equilibrium between business and user needs that JM clarifies through multi-channel visualisation and by including user insights provided fundamental information for workplace strategy and design. In the frog project the area-based blueprint developed allowed to unify branches and reduce their refurnishing resources by co-hosting more than one service per branch area.

13.5   Conclusions

Reflecting on differences observed between non-spatial and spatial journey mapping (Table 13.2) and insights from the case study we can say that JM has found some first applications for experiential spatial environments that include a diversity of service interactions (e.g. with the physical space, with representative of the service provider, with co-workers, with digital interfaces). Given that in Architecture and Real Estate management and corresponding industries JM is a less known and systematically applied method than in tech and digital service provider organisations and its representation differs from traditional journey maps (Table 13.2), it was worth trying to show an example of when JM is applied for workplace design and management.

The most interesting cultural shift that this method brings to workplace design and management is a shift of mindset from "workplace as an asset" to "workplace as a service," which helps to understand how the workplace operates in the context of the organisations' goals, and therefore become a tool for strategic decisions.

Continuing the parallel with the fields of product and service we can describe some long-term benefits of activating JM processes:

- Develop a mindset that allows workplace teams and their organisation to empathise with workplace actors and validate their needs and expectations, hence ensuring adoptions of new solutions and de-risking investments of resources for implementations;
- Leverage an evidence-based database of insights that forms a shared "source of truth" built on ongoing research, hence informing prioritisation, decision-making and road mapping to fix pain points and expand workplace functions and operations;
- Coordinate teams responsible for different phases of the experience, hence working strategically across organisational silos.

Adopting JM as a practice to design multi-channel workplace experiences would allow workplace design and management to improve the quality of workplaces-as-a-service. Moreover, embracing such a mindset encourages workplace researchers and professionals to look beyond the classic workplace users and also include external clients in the panel of key stakeholders (Davis, 2021). Indeed, workplaces are increasingly hybridising with a variety of complementary services and facilities (e.g., physical and mental well-being facilities, childcare facilities, workshop rooms, micro-production workshops, social spaces accessible by the external community, catering and private mobility services, etc.) where coworking spaces are the typical example of hybridisation (Migliore et al. 2021). Here not only the employees are using the workplace but other occasional users (among whom company clients and more) play a significant role as they enter the space and engage in an "experience" as well (Barbrook, 2006; Ricard, 2020). This chapter offered initial practical suggestions to expand the scope of JM from a tool to map experience to a system to design and manage workplace models holistically.

The articulation of all phases of JM, starting from generating a grounded understanding of the project scope and conducting dedicated user research to understand the current reality of users and workers deeply, allow organisations to develop what Levin defines conscious workplaces (Levin, 2005): workplaces whom settings, social interactions and supported processes are functional to reach their organisation strategic goals. The organisations that enable JM processes and respond to the resulting emerging needs can expect better workforce performances, talent retention, and cost optimisations.

Finally, a consideration on the capacity needed to adopt JM for the workplace is that its application requires the presence of human-centred design experts in support of architects and managers. Therefore, it configures as an accurate transdisciplinary method. For this reason, further research should explore possibilities for the development of hybrid figures internal to the organisation, able to leverage relations between JM and other existing methods, in particular, quantitative ones, specific to workplace studies and further explore specificities and more advanced representation techniques of this method for the built environment, starting from differences investigated in this chapter.

Acknowledgement

Contributor: Sara Manzini, Principal Service Designer, frog

13.6   Further Reading

Dermot, E. (2013). *Codesigning Spaces,* Artifice.
This book describes how to apply co-design methods to design successful collaborative workplaces.

Groves, K., & Marlow, O. (2021). *Spaces for Innovation: The Design and Science of Inspiring Environments*, Frame Publishers.
This book is the result of a research commissioned by innovation foundation Nesta. The book provides a framework for exploring the physical characteristics of spaces, illustrated through case studies of new generation of pioneering companies.

Penin, L. (2018). *An Introduction to Service Design: Designing the Invisible*, Bloomsbury Visual Arts.
This textbook on Service Design introduces a practical and theoretical overview of the fields of application of this discipline, with an analysis of the methods that include critical understanding of journey mapping.

References

Baird, L., Guiraud, F., Morgan, Z., & Sharpe, J. (2022). *Designing the Future of Work.* Frog. https://www.frog.co/designmind/designing-the-future-of-work
Barbrook, R. (2006). *The Class of the New*. Openmute.
Bridger, E., & Gennaway, B. (2021). *Employee experience by design: How to create an effective EX for competitive advantage.* Kogan Page.
Carbone, L. P., & Haeckel, S. H. (1994). Engineering customer experiences. Marketing management. *Marketing Management, 3*(3), 8–19.
Carlzon, J. (1987). *Moments of truth.* Ballinger Publishing Company.
Carrión, G. (2021). *The secret of growth? Reimagining employee experience*. Accenture. www.accenture.com/us-en/blogs/business-functions-blog/the-secret-of-growth-reimagining-employee-experience
Carter, C. (2022). *The secret language of maps: How to tell visual stories with data.* Stanford d.school Library, Ten Speed Press.
Chafi, M. B., & Cobaleda Cordero Cobelada, A. (2021). Methods for eliciting user experience insights in workplace studies: Spatial walkthroughs, experience curve mapping and card sorting. *Journal of Corporate Real Estate, 24*(1), 4–20. https://doi.org/10.1108/JCRE-12-2020-0069
Cobaleda-Cordero, A., & Babapour, M. (2022). Design for UX in flexible offices – Bringing research and practice together. In N.L., Black, W.P., Neumann, & I. Noy, (Eds.), *Proceedings of the 21st congress of the international ergonomics association (IEA 2021). IEA 2021. Lecture notes in networks and systems, vol 223* (pp. 3–11). Springer, Cham. https://doi.org/10.1007/978-3-030-74614-8_1

Davis, D. (2021). *5 Models for the Post-Pandemic Workplace*. Harvard Business Review. https://hbr.org/2021/06/5-models-for-the-post-pandemic-workplace

Deloitte. (2021). *The Digital Workplace Reimagined*. https://www.gartner.com/technology/media-products/newsletters/Deloitte/1-27HTLMDE/index.html

Downe, L. (2020). *Good services: How to design services that work*. Bis Publisher.

Fassi, D., Galluzzo, L., & De Rosa, A. (2018a, July). Service+Spatial design: Introducing the fundamentals of a transdisciplinary approach. In *ServDes2018. Service Design Proof of Concept, Proceedings of the ServDes. 2018 Conference, 18-20 June, Milano, Italy, 150*, 572–581. Linköping University Electronic Press.

Fassi, D., Galluzzo, L., & Marlow, O. (2018b). Experiencing and shaping: the relations between spatial and service design. In *ServDes2018. Service Design Proof of Concept, Proceedings of the ServDes. 2018 Conference, 18-20 June, Milano, Italy, 150*, 572–581. Linköping University Electronic Press.

Gallup. (2022). *Gallup's 2022 Guide to Employee Engagement*. www.gallup.com/corporate/212381/who-we-are.aspx

Groves, K., & Marlow, O. (2016). *Spaces for innovation: The design and science of inspiring environments*. Frame Publishers.

Kalbach, J. (2016). *Mapping experiences*. O'Reilly Media.

Korzybski, A. (1933). *Science and sanity. An introduction to non-Aristotelian systems and general semantics*. The International Non-Aristotelian Library Pub. Co.

Ku, B., & Lupton, E. (2022). *Health Design Thinking*, MIT Press.

Latour, B. (2005). *Reassembling the social: An introduction to actor-network-theory*. Oxford University Press.

Law, J., & Singleton, V. (2013). ANT and politics: Working in and on the world. *Qualitative Sociology, 36*(4), 485–502.

Leesman. (2018). *Workplace experience revolution*. https://www.leesmanindex.com/workplace-experience-revolution/

Levin, A. (2005). Changing the role of workplace design within the business organisation: A model for linking workplace design solutions to business strategies. *Journal of Facilities Management, 3*(4), 299–311.

Lichaw, D. (2016). *The User's journey: Storymapping products that people love*. Rosenfeld Media.

Manca, C., Grijalvo, M., Palacios, M., & Kaulio, M. (2018). Collaborative workplaces for innovation in service companies: Barriers and enablers for supporting new ways of working. *Service Business Journal, 12*, 525–550. https://doi.org/10.1007/s11628-017-0359-0

Marlow, O., & Egan, D. (2013). *Codesigning space: A primer*. Artifice Books on Architecture.

Migliore, A., Manzini Ceinar, I., & Tagliaro, C. (2021). *Beyond coworking: From flexible to hybrid spaces. Chapter of Human Resource Management book series (HRM)*. Springer.

Oxford English Dictionary (2013) 7th edition, Oxford Languages.

Ricard, S. (2020, December 10). *The Year of The Knowledge Worker*. Forbes. https://www.forbes.com/sites/forbestechcouncil/2020/12/10/the-year-of-the-knowledge-worker/?sh=92a9fba7fbba

Sanders, E., & Stappers, P. (2012). *Convivial toolbox: Generative research for the front end of design*. Bis Publisher.

Stickdorn, M., Hormess, M., Lawrence, A., & Schneider, J. (2018). *This is service design doing: Applying service design thinking in the real world*. Reilly Media.

Whitter, B., & Bersin, J. (2019). *Employee experience: Develop a happy, productive and supported workforce for exceptional individual and business performance*. Kogan Page.

Wride, M., Maylett, T., & Patterson, K. (2017). *The employee experience: How to attract talent, retain top performers, and drive results*. Wiley.

14 Compendium

Choosing methods for workplace research and management

Mardelle McCuskey Shepley

Cornell University, USA

Recent shifts in workplace culture have stimulated researchers to examine the implications of these transitions. One of the significant changes is the cultural/operational segue from the analogue to the digital workplace (Meske & Junglas, 2021) and the subsequent need to customize these workplaces to support the new activities. The need to study these co-located and satellite spaces' social and physical environmental characteristics was amplified by the recent pandemic during which many workers shifted to home environments to conduct business. This phenomenon was demonstrated by the multiple studies on the impact of dispersed locations on workplace culture and environment that were published between 2020 and 2022 (e.g., de Lucas Ancillo et al., 2021; Van Der Lippe & Lippényi, 2020; Vyas & Butakhieo, 2021) and futurists are suggesting that hybrid workplace environments may become the status quo (Iqbal et al., 2021; Williamson & Colley, 2022)

Workplace research has also been encouraged by a growing interest in the impact of the built environment on health (e.g., Laddu et al., 2021). Researchers have demonstrated the impact of lighting, nutrition, acoustics, air quality, ergonomics, and other environmental dimensions on human wellness in workplace settings. Organizations such as the WELL Building Institute have developed extensive guidelines (International Well Building Institute, 2016), backed by evidence, to help guide workplace protocols and guidelines.

These three topics: the transition from analogue to digital work tasks, hybrid work settings, and healthy work environments can be addressed by a variety of methodologies as this book gave evidence.

This chapter brings up some valuable elements that need to be considered when choosing methods. Although the research topic and focus are obviously one of the first aspects to consider, selecting the most appropriate method can be influenced by several factors that strongly affect the research design and its chances for success. The following sections collect some key hints and provide a brief guideline to avoid falling in love with one method and ending up with failure.

This chapter addresses three aspects of the research process. In order to initiate a study, we need to begin with establishing the parameters. What is the question we are trying to address? What have previous researchers done regarding the topic and how strong is the predecessor research? What are the potential operational and physical environmental variables and what are the expected outcomes? The second

DOI: 10.1201/9781003289845-14

aspect addressed is the criteria we use for selecting a method. What are the skills of the researcher and the research team? Is there an opportunity for a pilot study? What are the restrictions regarding human subjects' oversight groups (Institutional Review Boards (IRBs))? Lastly, what are the types of methods available to address workplace research?

14.1 Establishing research parameters

14.1.1 Questions and hypotheses

The first step for any researcher is to pose the questions or hypotheses that need to be addressed. As John Dewey noted, "a problem well put is half-solved" (Dewey, 1938, p. 108). Most hypotheses are a natural outcome of the questions that are posed and might be generated in response to comparing changing conditions (e.g., the impact of productivity of office versus remote working conditions), reacting to pressing research topics (e.g., the relationship between open and private offices and frequency of social interaction), or the need to confirm the impact of untested innovations (e.g., the impact of access to nature on worker productivity).

Hypotheses are likely to be generated by reviewing existing literature, which helps to determine where a substantive body of research is available and where new research is needed. Due to advancements in literature search platforms, systematic and scoping reviews have become very popular. Potential references can be culled using the SUPPORT system, which offers ways to evaluate the quality of a study (Hamilton & Shepley, 2009) (see Figure 14.1). Studies, whether qualitative or quantitative, should have a clearly defined method that could be replicated and potentially triangulated with other data. The research question or hypothesis should be communicated and contribute to new knowledge. The researchers should approach the subject with objectivity and, when publishing, subject to peer review.

Hypotheses encompass a subject population (e.g., staff) and independent and dependent variables. There are multiple independent and dependent variables associated with workplace hypotheses. In studies about physical space, the environment serves as the independent variable and in studies examining the impact of social/operational variables, protocols and human interactions serve as the independent variables.

14.1.2 Criteria that impact

Once a researcher is clear about his/her questions/hypothesis, potential methods reveal themselves. Criteria that might impact one's choice include: the cognitive style of the researcher, availability of precedent and pilot studies, research team strengths, availability of resources (funding, labour, etc.), and impact of the IRB process.

14.1.2.1 Cognitive style

Some researchers are more comfortable with qualitative methods, some with quantitative methods, and some prefer multi-methods. Regardless of one's preference,

S	U	P	P	O	R	T
Uses Substantive methodology	**Uncovers new knowledge**	**Is it Peer-reviewed?**	**Identifies Problem or Hypothesis**	**Has Objective researchers**	**Uses Repeatable methodology**	**Can the outcomes be Triangulated?**
The methodology must be examined to *ensure the credibility*.	Does the article describe research that has successfully *uncovered new knowledge*?	*Single-blind reviews* allow the reviewer to know the name of the submitter, but the author does not know the name of the reviewer(s).	Researchers should evaluate whether the article *clearly identifies the problem or hypothesis*.	Researchers are cautioned to *maintain their objectivity*.	The methods should be *sufficiently clear* so the study can be replicated.	A *multi-method approach* is not required but might be helpful.
Use *random sampling*	Address a new idea or build on an old one.	A *double-blind review* is one in which the reviewers are not revealed to the authors and the authors are not revealed to the reviewers.	The intentions of the researcher need to be articulated in the *introduction* to a paper.	Research on a firm's projects can be conducted by members of the firm. However, caution must be exercised to *avoid contaminating results*.	The outcomes *can be retested by others* to build a body of knowledge around a specific question.	Triangulated outcomes are the result of multiple methodologies applied to a single study question.
Have a high *questionnaire return rate*	*Ties articles together* in a new way				Studies that cannot be replicated, either from lack of information or lack of support of the hypotheses, eventually fall by the wayside.	Examples: interviews, behavioral observations, and questionnaires.
Review the *reliability and validity* of the instruments	*Identify holes* in the body of research					Research will often use a *combination of qualitative and quantitative approaches* to gather a full understanding of a problem.
	Set the information in a historical or conceptual *context*.					

Figure 14.1 Support system for evaluating papers (Hamilton & Shepley, 2009)

qualitative methods may be needed to develop a grounded theory that establishes research questions if a topic has not been explored previously. A study does not have to be quantitative to be reliable and valid. Qualitative studies can be substantive if carried out in a structured manner using previously vetted methods. Naturalistic inquiry, for example, which statisticians developed, involves an orderly analysis of interview and observation data using auditing, member checking, and case reporting techniques (Lincoln & Guba, 1985). While qualitative research analysis may be more subject to researcher bias; engaging external team members can help neutralize this effect. Consulting with outside experts, incorporating a broad demographic of participants, or using purposive samples are some of the factors that strengthen qualitative research.

Many believe multi-methods are necessary to understand socio-environmental issues. Multi-methods look at a question from various perspectives or use one method to provide information for a subsequent study phase. Designers tend to prefer a multi-method approach, as their training is multidisciplinary (i.e., incorporates both aesthetics and technology).

14.1.2.2 Research team strengths

Another factor to consider when selecting a method is the skill set of the research team and the team's consultants. Some investigators are good at inductive reasoning (inferring general ideas from particular events) and others are deductive thinkers (inferring particular events from general ideas). A strong team includes both during the hypothesis development and data interpretation phases. The activities of statisticians are not limited to analyzing data. Statistical consultants can recommend ways to gather data and reflect on the implications of different approaches and conclusions.

14.1.2.3 Precedent and pilot studies

Both precedent studies and preliminary pilot studies are essential. Precedents (previous studies) can serve to validate tools and avoid the additional work of customizing research methods. A pilot study provides the opportunity to vet the process and tools and/or provide data that will support the design of subsequent portions of the study. Regardless of the perceived accuracy of the proposed research tools, unforeseen problems often occur, and pilot studies help to reduce future errors.

14.1.2.4 Resources

The availability of funding and space resources may also influence the potential methods as different study processes have pros and cons in this regard.

Paper surveys support participant anonymity but may be more expensive, environmentally detrimental and, unless computer-readable, data must be transferred by hand, which creates the opportunity for error. Online surveys are less expensive, and the data can be uploaded directly, but may require the purchase of software and may reveal participants through email information. Behavioural observation gives a clear sense of human activity but is personnel intensive and time-consuming. Physiological measures are readily quantifiable, but often require expensive medical equipment and trained staff.

Spatial resources may also be an issue if research space is needed for interviews and focus groups, gathering of biomarkers, equipment storage, or other activities.

14.1.2.5 Human subjects review

The impact of an IRB should also be considered. When academics study the workplace, human subjects' approval will be required and, in certain circumstances (i.e., healthcare work settings), the process can delay a project for three months to a year, depending on the vulnerability of the study participants. For other workplace environments, approval may not be as difficult, as office organizations rarely have human subject review boards. If information is needed quickly, the researcher may opt to use less intrusive measures (anonymous questionnaires as opposed to personal interviews) to facilitate the completion of the study.

14.2   Types of methods

Environmental data might be gathered (e.g., energy consumption, acoustical, and light levels), but workplace methods' research often focuses on human behavioural outcomes. "Behaviour is what an organism is doing at any moment […]. In environment behaviour research […] the behaviour must always be seen within an environmental context" (Bechtel & Zeisel, 1987, p. 12) and can include a variety of methods. The most commonly used methodologies and associated tools are listed in Table 14.1. Many of them are described in detail in the book chapters.

Among the *methodologies* that are discussed in-depth in the previous chapters, there are Critical Discourse Analysis (CDA), Delphi studies, Ethnography, and Social Network Analysis (SNA). CDA, Delphi studies, Ethnography, and SNA are processes that manipulate data typically received through self-report and observation.

14.2.1 Archival methods

Archives provide useful census data, employee data, GIS data, and similar regularly collected information. Archival information that might benefit workplace studies includes: employee retention data, sick leave data, and recruitment data. Census data regarding demographics at a regional scale might be useful in identifying appropriate cultural and physical environmental features for potential employees. GIS data can be used to survey amenities at a neighbourhood scale. Employee data will aid in understanding the impact of change within an organization.

14.2.2 Behavioural observation

Observation may be more accurate than participant self-report (Pryor et al., 1977) regarding the impact of a physical environment or an organizational shift. However,

Table 14.1 Workplace methods and associated tools (those with reference are described in the other chapters of this book)

Methodologies	Tools
Archival methods	
Behavioural observation	Direct or indirect
	Photography
	Video recording
Critical discourse analysis (**Chapter 5**)	
Delphi studies (**Chapter 9**)	
Ethnography (affective, auto-, and digital) (**Chapters 2, 3, and 4**)	
Social network analysis (**Chapter 10**)	
Self-report	Adjectival checklists
	Cluster analysis (**Chapter 7**)
	Cognitive (journey) mapping (**Chapter 13**)
	Diary studies (**Chapter 6**)
	Focus groups
	Interviews
	Photovoice
	Psychological tests
	Questionnaires (**Chapter 11**)
	Stated preference studies (**Chapter 8**)
Spatial analysis	Physical traces
	Spatial walk-through
	Space syntax (**Chapter 12**)

the method has shortcomings, such as the difficulty of recording a sequence of activities without missing simultaneous behaviours (Canter, 1972). To account for this, a researcher may have to limit the number of behaviours or have simultaneous observers. There are also challenges regarding interpreting behaviour and the impact of unexpected environmental events such as changes in weather or holidays. Three topics should be considered when developing an observational method: degree of researcher engagement, recording device, and what to observe.

- *Degree of engagement.* There are four levels of engagement: clandestine observer, acknowledged observer, marginal participant, and engaged participant. Participants might alter their behaviours when they feel they are being observed. To reduce this effect, a researcher might use unobtrusive measures (which may raise ethical issues), reduce conspicuousness through their location, or engage in extended sessions that permit adaption to the observers' presence.
- *Recording device.* Behaviour can be recorded via a checklist, hand notation, photographs, maps, video, or computerized notation. Missed information is particularly common when recording behaviour on paper, due to the rapidity with which data must be gathered. Technology, such as tablets programmed with observational software, has enabled researchers to record behaviours directly and quickly. Video recordings provide the opportunity to control for the subjectivity of the observer by allowing multiple subsequent reviews.

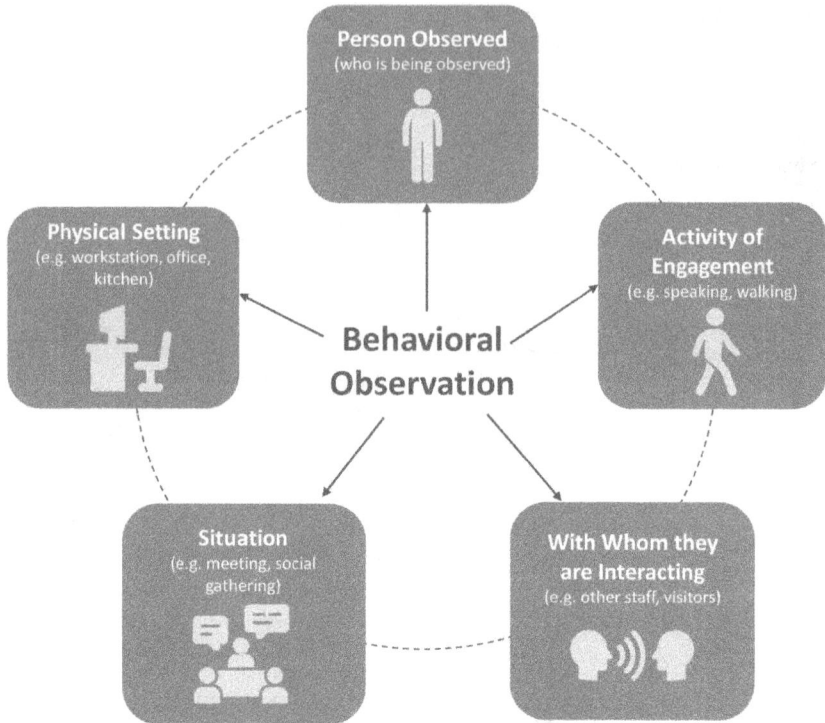

Figure 14.2 Behavioral observation components

- *What to observe*. Apart from the individual(s) being observed and the physical setting, behavioural observation might include the activity in which they are engaged (e.g., speaking and walking), duration and frequency of behaviours, the persons with whom they are interacting (e.g., other staff, visitors), and the context in which they are situated (e.g., a meeting and a social gathering) (See Figure 14.2).

14.2.3 Self-reports

Self-reports are: gaming, forced decision, anecdotal/experiential, psychological tests, graphic responses, questionnaires, interviews, cognitive mapping (subjects draw maps of perceptions), user diaries (to record periodic observations), psychological tests, and adjectival checklists. This book presents a sample of the *tools* (see Table 14.1) that support self-reports, often in combination with each other, including cluster analysis, cognitive (journey) mapping, diary studies, questionnaires, and stated preference studies.

Self-reports may have challenges regarding validity; they provide a more personal approach to participant engagement. Other advantages are: the opportunity they present to uncover experiential definitions, covert data, and retrospective data.

On the other hand, self-reports may lack experiential realism due to the subject awareness of the researcher and participants may be impacted by the dissonance reduction phenomenon. Other shortcomings are the restrictions due to lack of participant cooperation, the requirement that the subjects (e.g., young children) have a specific level of mental functioning, and participant fatigue. Self-reports often need a large number of respondents to be meaningful.

A survey measures the opinions or experiences of people by asking questions. Among the issues associate with a survey questionnaire are its length, whether it is open versus directed; the order of the questions; the use of graphics; measurement scales; and which demographic information should be solicited. Gender and race/ethnicity are particularly challenging. If a sample size is too small to reflect the complexion of the larger population, then some questions might not be useful. The questionnaire, interview, and focus group protocols can be: structured (gather succinct responses), open-ended (flexible, typically narrative), or a combination. Researchers tend to prefer the former when addressing a specific hypothesis and the latter useful when conducting research to determine potential hypotheses. Focus groups have the benefit of allowing for more in-depth responses, the interaction between participants, and non-verbal communication (Nagle & Williams, 2013).

Photovoice (a method whereby participants are invited to photograph environmental elements representing specific issues) allows participants to drive research questions. Researchers have engaged in processes similar to photovoice for decades (e.g., Aitken & Wingate, 1993; Bensen, 1995), and the process was first detailed by Wang and Burris (1997). Photovoice is particularly appropriate for marginalized populations; based on "empowerment education, critical consciousness, feminist theory, and documentary photography" (Sutton-Brown, 2014, p. 169).

14.2.4 Spatial analysis

Examples of spatial analysis are physical traces, space walk-throughs, and space syntax. Physical traces measure: effects of use (wear of physical features in the environment, such as flooring or furniture), changes in use (changing a staff lounge into a private workspace), and accretion (graffiti and public messages such as ad hoc signage in support of wayfinding). Space walk-throughs are environmental tours that reveal behaviour and space use. Space syntax, to which one of the chapters of this book is dedicated, is a *tool* that analyzes specifically spatial configuration.

14.3   Future of workplace research methodologies

Several trends in workplace research are likely. We can predict that work-at-home and other shared spaces, like coworking spaces, will expand in number and workplace research will advance extant social science and design research methods. The science is relatively new and is developing the most effective protocols. Related to this is the probability that social science and design researchers' goals will increase in their overlap. The interdisciplinary nature of this collaboration will contribute to the development of standardized methods.

Another likely trend is that simulation studies will become more common as the technology continues to refine itself. It is now commonplace for researchers to avail themselves of virtual reality and augmented reality to explore ideas. Related to advances in technology is the growing role of "big data" which may play a more significant role when examining workplace phenomena at the regional scale.

Lastly, physical and mental health are prominent global concerns, and wellness in the workplace will continue to be a strong research focus. Workplace researchers will place greater emphasis on increasing diversity in the subject populations studied and the construction of studies that reflect a broader range of cultural needs.

Sociologists, organizational experts, and designers are harbingers of a social perspective that acknowledges the impact of our workplace environments on productivity, satisfaction, and well-being. The papers that precede this chapter provide insights into how this type of research might be most effectively pursued.

References

Aitken, S. C., & Wingate, J. (1993). A preliminary study of the self-directed photography of middle-class, homeless, and mobility-impaired children. *The Professional Geographer*, *45*(1), 65–72.

Bechtel, R. B., & Zeisel, J. (1987). Observation: The world under a glass. In R. B. Bechtel, R. W. Marans, & W. Michelson (Eds.), *Methods in environmental and behavioral research* (pp. 11–40). VNR Company.

Bensen, L. (1995). *Criteria for designing leisure spaces for adolescent* [Masters thesis], Texas A&M University.

Canter, D. (1972, September). Royal hospital for sick children. *The Architects Journal, 174*, 527–569.

de Lucas Ancillo, A., del Val Núñez, M. T., & Gavrila, S. G. (2021). Workplace change within the COVID-19 context: A grounded theory approach. *Economic Research-Ekonomska Istraživanja, 34*(1), 2297–2316.

Dewey, J. (1938). *Logic: Theory of inquiry logic*. H.H. Holt and Co.

Hamilton, D. K., & Shepley, M. M. (2009). *Design for critical care*. Routledge.

International Well Building Institute (IWBI). (2016). *The WELL building standard*. https://standard.wellcertified.com/sites/default/files/The%20WELL%20Building%20Standard%20v1%20with%20May%202016%20addenda.pdf

Iqbal, K. M. J., Khalid, F., & Barykin, S. Y. (2021). Hybrid workplace: The future of work. In *Handbook of research on future opportunities for technology management education* (pp. 28–48). IGI Global.

Laddu, D., Paluch, A. E., & LaMonte, M. J. (2021). The role of the built environment in promoting movement and physical activity across the lifespan: Implications for public health. *Progress in Cardiovascular Diseases, 64*, 33–40.

Lincoln, Y. S., & Guba, E. G. (1985). *Naturalistic inquiry*. Sage.

Meske, C., & Junglas, I. (2021). Investigating the elicitation of employees' support towards digital workplace transformation. *Behaviour & Information Technology, 40*(11), 1120–1136.

Nagle, B., & Williams, N. (2013, September 12). *Methodology brief: introduction to the focus group*. Center for Assessment, Planning and Accountability. http://www.mmgconnect.com/projects/userfiles/file/focusgroupbrief.pdf

Pryor, J. B., Gibbons, F. X., Wicklund, R. A., Fazio, R. H., & Hood, R. (1977). Self-focused attention and self-report validity. *Journal of Personality*, *45*(4), 513–527.

Shepley, M. (2011). *Health facility evaluation for design practitioners*. Aesclepion.

Sutton-Brown, C. A. (2014). Photovoice: A methodological guide. *Photography and Culture*, *7*(2), 169–185.

Van Der Lippe, T., & Lippényi, Z. (2020). Co-workers working from home and individual and team performance. *New Technology, Work and Employment*, *35*(1), 60–79.

Vyas, L., & Butakhieo, N. (2021). The impact of working from home during COVID-19 on work and life domains: An exploratory study on Hong Kong. *Policy Design and Practice*, *4*(1), 59–76.

Wang, C., & Burris, M. A. (1997). Photovoice: Concept, methodology, and use for participatory needs assessment. *Health Education & Behavior*, *24*(3), 369–387.

Williamson, S., & Colley, L. (2022, February). *Working during the pandemic: The future of work is hybrid*. UNSW Canberra Public Service Research Group and CQUniversity. https://www.unsw.adfa.edu.au/sites/default/files/documents/Working_during_the_pandemic_the_future_of_work_is_hybrid_Feb_2022.pdf

15 Wind-up

On methodological takeaways and the importance of workplace research

Marko Orel

Prague University of Economics and Business, Czech Republic

"The world is not a solid continent of facts sprinkled by a few lakes of uncertainties, but a vast ocean of uncertainties speckled by a few islands of calibrated and stabilized forms".

– Bruno Latour

Every (hand)book has a story to tell and wisdom to share. The late Bruno Latour's (2005, p. 245) rather resonating quote from his work on actor–network theory captures the idea behind sharing scientific approaches and methodological forms that can tackle identified research subjects and problems. Thinking first of craftsmen and their manual labour, and later of clerks and administrative tasks, places of work have always been complex. With work transforming from the fairly rudimentary "sweat of one's brow" to a set of coordinated assignments that involved various processes and unrelated methods, the work environments have started to become multiplex social systems, where both physical and organizational design have begun to play a significant role in how these places affect individuals. They became focal points of interest for researchers from various fields and disciplines. And in parallel with growing engrossment, workplace research gradually became a research field that required a stabilized form of methodological approaches to allocated problems.

One thing is certain: contemporary workplaces are changing – and as we pointed out in the handbook's Preface, they are changing fast. Societal disruptions, such as the recent COVID-19 pandemic that prompted a (re)discovery of remote work practices, have temporarily made physical workplaces non-essential, accelerating the trend of workplace digitalization. While workplaces tend to follow the organizational narrative and managerial perspective, workers' behavioural patterns are also subject to transformation. Looking at the said alternations and the metamorphosis of the knowledge worker from a different perspective, contemporary workplaces prompt a boundless array of questions that workplace scholars attempt to address. The handbook overviewed a dozen methodological approaches that can accompany a practitioner or researcher eager to comprehend a chosen workplace site and dig deeper into its organizational cultures, languages, internal networks, physical settings, virtual processes, and other portraying elements.

DOI: 10.1201/9781003289845-15

Having said that, different methodological approaches can be used in distinctive settings and bring distinguishable data sets that are especially valuable when attempting to share light on a peculiar subject. Workplace autoethnography, for example, is especially suitable for research where the scholar tends to examine a particular work setting using their fellow participant's perspective. Since workplace autoethnography undertakes an analytical form and connects the data to find theoretical conclusions, it requires a rather systematic inquiry approach. While the latter requires more attention from the auto-ethnographer, it enables the establishment of a sense of authenticity by constructing complex narratives through personal and cultural experiences within a selected workplace.

While workplace autoethnography can be limited to some extent and potentially subjective, an affective ethnography could be utilized to comprehend a vast array of unalike workplace elements. In summary, an effective ethnographic approach acknowledges that the said elements, such as workplace language, involved individuals, accompanying texts, and relevant agencies, are entangled and should be viewed through one another as moving data. The acknowledgement of the latter is fundamental when an ethnographer is affected by varying elements during the fieldwork process, but nevertheless continues the study by recognizing the nature and intra-activity of the latter.

With the swift advancement of new technologies, many organizations complement their physical workplace settings with digitalized counterparts such as computer-mediated software or more detailed and immersive digital workplace twins that can be accessed using a selected extended reality technology. These gradual changes may prompt workplace scholars to use digital ethnography. Through computer-mediated social interactions, this exploratory approach uses ethnography to examine organizations, workplace-related cultures, and user groups. In other words, digital ethnography enables scholars to either replace or complement their exploration of online workplace components through a preferred technological tool.

When workplace researchers allocate certain problems, Critical Discourse analysis can be used to tackle pressing difficulties by analyzing the discourses that uncover and reanimate these stumbling blocks. Moreover, it enables scholars to vigorously assess the meaning of the language when the latter is used for describing and explaining. As such, the Critical Discourse analysis grants the ability to address workplace-related social problems through critically investigating issues of relevant language and texts, comprehend any appropriate internal relations that are discursive and overlined as powerful, and assist in understanding the organizational essence behind the selected workplace.

A workplace scholar can also use a diary study to expand on this and seek insights into workplace users' behaviour, interactions, and activities over time. The latter stands for a methodological approach that uses a set of assessment methods that enable a scholar to examine workplace and internal processes in real-time or observe users' behavioural patterns and experiences over a defined period. In other words, the diary methodology allows the researcher to record entries about workplace users' lives over specific actions and around peculiar actions.

Although these summarized qualitative methodological approaches can be extensive and explanatory in their nature, it is not necessarily sufficient to pinpoint the pressing issues that scholars can identify. Cluster Analysis, for example, is a data reduction technique commonly used in quantitative research to group observations into exclusive groups. While Cluster Analysis requires adequate statistical skills and relatively large amounts of data, it can enable a workplace researcher to group a set of objects so that objects in the same clusters become similar to one another compared to those in other groups. The latter especially comes in handy when studying multilocation work and workers' location choices.

Multilocation work and the preferential choice by a selected individual or group of workers can be further uncovered using Stated Choice experiments. Consider various data types collected through interviews, surveys, or other data-gathering techniques. Then, the results can provide quantitative measures of the relative importance of the attribute of the choice alternatives. These might include an individual's willingness to pay for a workspace, work from a noisy, café-like place, or prefer an open-plan work environment. Either way, the approach behind the Stated Choice experiments can provide workplace scholars the ability to predict the probability that specific alternatives will be selected.

Alongside the described pathway of tackling an identified workplace-related problem, a researcher might use the Delphi technique. The latter is a well-established methodological approach that enables scholars to answer a specific question by reaching stability in opinion across the subject experts. The approach comes especially handy when the knowledge available to a researcher is partially incomplete or subject to uncertainty. To tackle the latter, the Delphi approach builds on a technique of asking multiple questions to the selected target group of individuals considered experts in their respective fields or directly interlaced with the research subject.

Working individuals are habitually one of the best sources to gather data. A Social Network analysis approach stands for a methodological choice that studies the workers' social interactions and relations within the workplace. It can be used to comprehend both social and spatial workplace factors. Moreover, the Social Network analysis allows a researcher to investigate the relationships between a selected workspace and other related social characteristics.

To further explore the characteristics related to a workplace, a scholar can use the Post-occupancy Evaluation survey approach. The latter represents a methodological approach that grants the collection of information from a predefined population of workplace users, aiming to gain insights into the quality and usability topics that tackle an identified workplace-related problem. What is more, Post-occupancy Evaluations enable scholars to stand firm with the culture of evidence-based research design, mainly because they can communicate and use the obtained results better.

In line with the latter, it is vital to overview Space Syntax, a set of techniques that enable workplace scholars to analyze the spatial configuration of a selected site and gain relevant insights into human activity patterns. As the Space Syntax methodology builds on a human-focused approach to investigate relationships between a range of social, psychological, and economic factors, and a spatial design of a workplace, the approach can prove to be significantly crucial in swiftly changing workplace design and usability patterns. An additional step could be the parallel use of Journey Mapping, which allows scholars to visualize a workplace user's process to complete a work-related task.

These are only a handful of methodological approaches that enable researchers to dig deeper into the subjects related to contemporary places of work. The handbook that explored these undertakings on obtaining relevant data has been developed with great care by both editors and contributing authors and should withstand the test of time. The described methodological approaches have been repeatedly used across various other disciplines and grant access to commonly relevant data sets. With workplaces continuing to evolve and adapting to new forms of how the work is being conducted, we can go back to Latour's opening quote and draw a concluding parallel. Indeed, the fascinating world of workplaces alternates and brings a vast array of uncertainties. Still, stable, and sound methodological approaches can assist in tackling any pressing issues that one might inquire about. And for that, the handbook you are currently holding onto in either a paper or digital format can assist you along the way of your journey as a workplace scholar or workplace practitioner.

Reference

Latour, B. (2005). *Reassembling the social: An introduction to actor-network-theory*. Oxford University Press.

Index

Note: **Italicized** and **bold** page numbers refer to figures and table. Page numbers followed by "n" refer to notes.

For Product Safety Concerns and Information please contact our EU
representative GPSR@taylorandfrancis.com
Taylor & Francis Verlag GmbH, Kaufingerstraße 24, 80331 München, Germany

www.ingramcontent.com/pod-product-compliance
Lightning Source LLC
Chambersburg PA
CBHW060256220326
41598CB00027B/4120

9 7 8 1 0 3 2 2 6 7 6 8 5